conran DESIGN *guides*

LAMPS AND LIGHTING

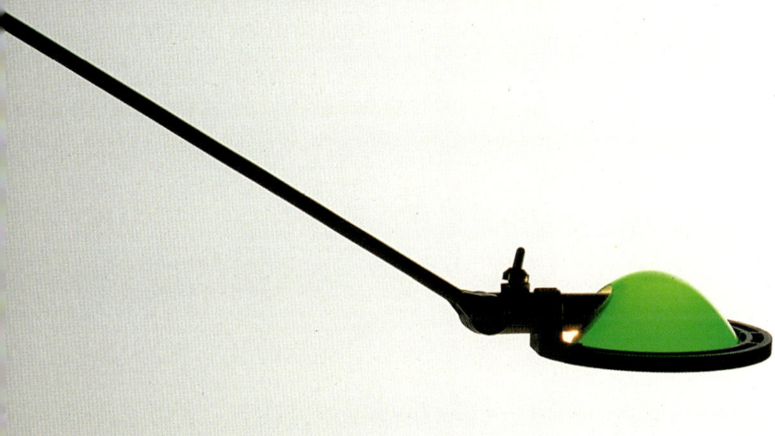

JEREMY MYERSON
& SYLVIA KATZ

VNR VAN NOSTRAND REINHOLD
New York

Series Editor Joanna Bradshaw
Editor Mary Davies
Editorial Assistant Sally Poole
Design Paul Welti
Illustrator Cherrill Parris
Picture Research Nadine Bazar
Picture Research
Assistant Gareth Jones
Production Sonya Sibbons

16 15 14 13 12 11 10 9 8 7 6 5 4 3 2 1

Library of Congress Cataloging-in-Publication Data

Conran, Terence.
The Terence Conran design guides–lamps and lighting / edited by Sir Terence Conran.
p. cm.
ISBN 0-442-30302-5
1. Electric lamps, Incandescent–History. 2. Design, Industrial.
I. Title.
TK4351.C66 1990
749'.63–dc20 89-70663
CIP

Published in the United States of America by
Van Nostrand Reinhold
115 Fifth Avenue
New York, New York 10003

First published in Great Britain in 1990 by
Conran Octopus Limited

Typeset by
Servis Filmsetting Limited
Printed by Wing King Tong

ACKNOWLEDGMENTS

The publisher thanks the following photographers and organizations for their kind permission to reproduce the photographs in this book:
5 top Flos; 5 above Louis Poulsen; 5 below Flos; 5 bottom Artemide; 6 Philips Lighting; 7 Ann Ronan Picture Library; 8–9 Philips Lighting; 11 Optelma Lighting; 16 David A Hanks & Associates; 18 Artemide; 19 One-off, London; 22–23 Flos; 24 left Sotheby's London; 24 right courtesy J Alastair Duncan; 25 Angelo Hornak (Editions Graphiques, London); 26 left Sotheby's London; 26 centre Sotheby's London; 26 right CCI, Paris; 27 The Metropolitan Museum of Art, Purchase, Theodore R Gamble, Jr. Gift in honour of his mother Theodore R Gamble 1982 (1982.33); 28 left Sotheby's London; 28 right Bauhaus-Archiv, Berlin (photograph by Lepkowski); 29 left Sotheby's London © DACS, London 1990; 29 right BIP Ltd; 30 Artemide; 30–31 Aldo Ballo; 31 Flos; 32 left Artemide; 32 right Verner Panton; 33 left O-Luce Italia SpA IV from *Design Since 1945*; 33 right CCI, Paris; 34–35 Louis Poulsen; 36 Sotheby's London; 37 Henning Rogge (from "Industriekultur: Peter Behrens und die AEG" 1907–1914) © DACS, London 1990; 38 left By courtesy of the Board of Trustees of the Victoria and Albert Museum © DACS, London 1990; 38 right Danish Design Centre; 39 left Philadelphia Museum of Art: Purchased; 39 right Danish Design Centre; 40 right Flos; 40 left Louis Poulsen; 41 Flos; 42 left Artemide; 42 right Verner Panton; 43 Ingo Maurer; 44 left Flos; 44 right Artemide Litech; 45 above Erco Lighting Ltd; 46–47 Flos; 48 left Bauhaus-Archiv, Berlin; 48 right Clive Corless/Conran Octopus; 49 Anglepoise Lighting; 50 left Thousand & One Lamps Limited; 50 right Walter Dorwin Teague Associates Inc; 51 left courtesy of Wolsonian Foundation, Miami, Florida; 51 right courtesy of Wolsonian Foundation, Miami, Florida; 52 left CCI, Paris; 52 right O-Luce, Italian, SpA; Clive Corless/Conran Octopus; 54 Flos; 55 above CCI, Paris/Artemide; 55 below Crescent; 56 Stilnovo; 57 David Morgan Associates; 58 Artemide; 58–59 Artemide; 60 left Sotheby's London; 60 right Nicola Grifoni/Conran Octopus; 61 Artek Oy; 62 Flos; 63 left Aldo Ballo; 63 right DP, Milan; 64 above Jane Dillon; 64 below Sotheby's London; 65 left Flos; 65 right Flos; 66 Aldo Ballo; 67 left Sottsass Associati; 67 right Memphis Milano; 68 left Tony Wills; 68 right Ecart International; 69 left Flos; 69 right Artemide; 70 Ingo Maurer; 70–71 Candell Lighting; 71 One-Off, London; 72–73 Artemide; 74 left Flos; 74 right DP, Milan; 76 Danish Design Centre; 77 Verner Panton; 78 Flos; 79 Stylograph (Jean-Pierre Godeaut).

The following photographs were taken specially for Conran Octopus by Simon Lee: 2–3, 45 below, 65 centre.

We would like to thank the following for their cooperation:
Atrium Ltd, London
Design Museum
London Lighting Company

Every effort has been made to trace the copyright holders and we apologize in advance for any unintentional omission and would be pleased to insert the appropriate acknowledgment in any subsequent edition of this publication.

AUTHORS' ACKNOWLEDGMENTS
The authors wish to thank all those manufacturers and designers who answered queries and searched through their archives, the supportive and professional staff at Conran Octopus and Sir Terence Conran for his personal interest and guidance.

NOTE TO READER
Names of objects and designers printed in roman or **bold** type denote that a photograph of the object or a biography of the designer can be found elsewhere in the book.

introduction

THE IMPORTANCE OF LIGHTING DESIGN

Light fascinated artists, architects and engineers long before the invention of modern electric lighting gave them a new, unrivalled technological palette to explore in the twentieth century. Awe-struck appreciation of light's ability to define colour, texture and form, and its capacity to shape emotional response can be found in the great art, and literature, of even the earliest civilizations. From shooting stars, volcanoes and lighting to fireworks, wax candles and gas lamps – the central role of light in visual perception can be traced through the culture of centuries.

It was not until a series of technological breakthroughs in the late nineteenth century led to the creation of the modern tungsten-filament electric light bulb that scientists and engineers could really begin to control, direct and diffuse light of their own design.

Importantly, the development of modern lighting coincided with the emergence of the profession of industrial designer in the early years of the twentieth century. Design pioneers in Europe and the USA immediately turned their attention to the challenge of lighting by addressing the design of the light

Early demonstration of an arc lamp at the Place de la Concorde in Paris, 1844.

fitting – the instrument which controls, directs and diffuses the light beam in any given space.

The light fitting or fixture became one of the most important symbolic objects by which advances in design thinking and practice were measured. That still holds true today. Like the chair or dinner set, the light fitting presents a challenge to designers as an essential commodity for millions of people. It is also constantly changing as materials technology and scientific research advance.

Lighting is one of the most difficult components of interior and industrial design to get right. Its principles involve a complex combination of art and science, technical precision and emotional perception. That is why the subject has produced such restless design innovation decade after decade. The continual process of improvement goes on today.

This book surveys the milestone developments and classic designs of modern lighting. But because of the special place the modern light fitting has enjoyed in the vanguard of design exploration, it is more than a buyer's guide to the great designs: it reflects the influential design movements and ideologies which have shaped the material world of the twentieth century – from Art Nouveau, the Bauhaus and the Machine Age to Scandinavian Modern, Memphis and Post-Modernism.

Thomas Edison's carbon-filament incandescent bulb, as explained in the *Scientific American*, 1880.

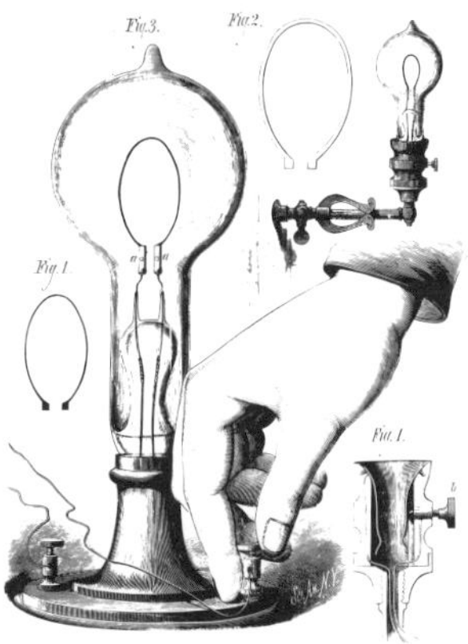

TECHNOLOGICAL ADVANCES

To understand the pattern of design development in lighting in this century, it is necessary to know in broad outline the technology on which it has been based.

Thomas Edison, the great American inventor, is the first, and certainly best-known, figure in this story. 'We will make the electric light so cheap that only the rich will be able to buy candles,' he exclaimed shortly after conducting an experiment in October 1879 at Menlo Park, New Jersey, with a device

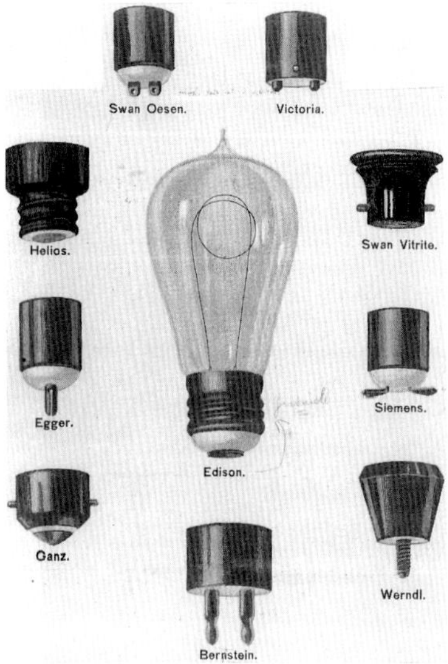

Selection of lamp caps showing the lack of standardization in the early days of manufacture.

which was the prototype of the tungsten–filament bulb.

Edison was a tireless inventor who took out no less than 1,093 patents during his lifetime for devices including the electric pen, the phonograph, waxed paper, and, of course, the incandescent electric lamp. In 1878 he had formed the Edison Electric Light Company, forerunner of General Electric, and conducted research into ways to develop a lamp or bulb with a resistant filament. Early attempts burnt out but, when he replaced aluminium filaments with carbon ones the following year, he achieved a breakthrough. The first commercially viable electric lamp, then, comprised a carbon-wire filament in a glass bulb from which all air had been evacuated; the ends of the wire were drawn through a sealed cap to the electric supply. When the supply was connected, the filament glowed, providing visible light. The vacuum within the bulb slowed down the process of oxidation which ultimately destroyed the filament. When Edison's achievement was heralded on the front page of The New York Herald *in December 1879, gas stocks and shares plummeted on Wall Street.*

However, a carbon-filament lamp was also being created on the other side of the Atlantic at the same time. In 1878 the English inventor Joseph Swan conducted an experiment which heated the filament while evacuating gases from a vacuum-filled glass envelope. By 1880 the first domestic lighting system using filament bulbs was seen in Britain; they were installed at the fashionable new Northumbrian home of British armaments millionaire Lord Armstrong.

Edison and Swan sensibly decided to join forces. They formed the Edison and Swan United Lamp Company, later known as Ediswan. Each had invented his own method of fixing the light bulb to the supply circuit: Edison patented the screw-in brass cap; Swan devised a bayonet fitting. Both are in use today.

But although the carbon-filament bulb was a significant advance on the oil and gas lighting of the nineteenth century, it lacked durability and efficiency. Research focused on improving the filament, and a number of metallized alternatives emerged. Then, in 1908, William Coolidge, a New York scientist with the General Electric Research Laboratory, found a solution. He produced commercially practical tungsten filaments which could be drawn to specific diameters and which were rugged enough to withstand the shocks and vibrations of the street and railroad cars of the period.

In 1913 Coolidge's associate, Irving Langmuir, solved the problem of the blackening of the bulb as the tungsten glowed. It was caused by early tungsten- and carbon-filament wire alike when molecules of filament migrated to the bulb, reducing efficiency. The introduction of a little inert gas (at first, nitrogen and later a mixture of argon and nitrogen) reduced migration, giving whiter light and longer life. Later, in 1933, the introduction of the double-coiled filament significantly increased bulb life.

The important pioneering work on the tungsten-filament bulb, mainstay of twentieth-century domestic lighting, was complete. Despite a continual process of innovation and updating, the next major breakthrough in the field did not come until 1959 when the first practical tungsten-halogen lamp was developed in the USA by Zuber and Moser.

Advertisement for the Dutch lamp-maker Philips, 1913, stresses the strength of drawn tungsten filament.

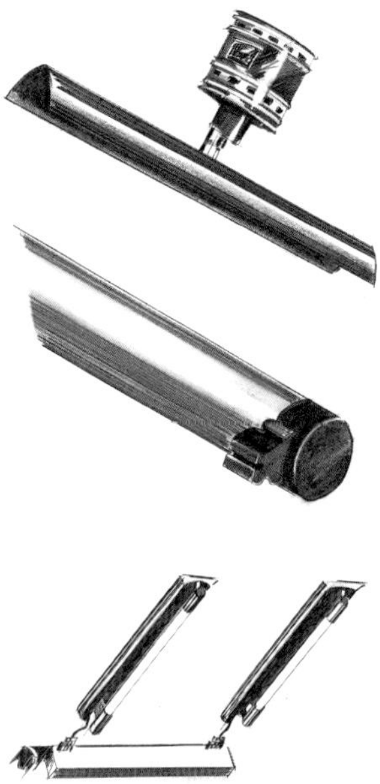

Fluorescent lighting has developed dramatically from early pioneering efforts (*above*) to the modern range of mini-fluorescents (*below*) which can slot into small domestic fittings.

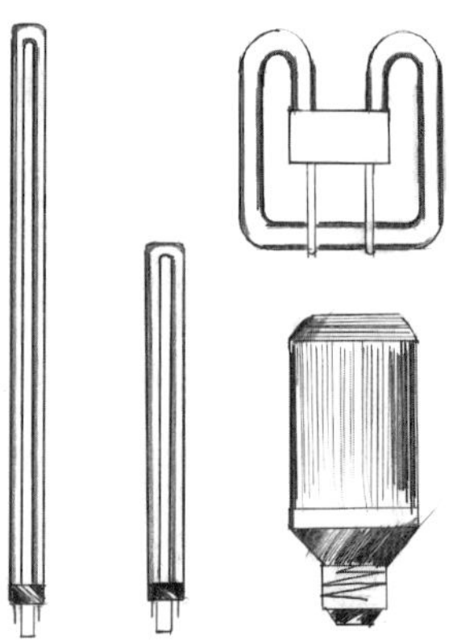

Meanwhile the big rival to the tungsten-filament bulb in the home, the fluorescent tube, was not introduced until 1938. But when General Electric launched it on the American market, it had an immediate impact: it was more energy efficient, longer lasting and cooler than tungsten. This was because it radiated light in a different way. The fluorescent lamp comprised a glass tube coated with phosphors and filled with a mixture of argon and mercury vapour. Metal electrodes were positioned at either end and when current flowed through the gas between them ultraviolet radiation was produced. The phosphor coating absorbed the radiation and fluoresced (or 're-radiated' energy as visible light).

Fluorescent created a host of new opportunities for lighting designers but its light was seen in some quarters as cold and unsympathetic. It was regarded as ideal for offices or factories, and for functional areas of the home, such as kitchens or garages. But it lacked the warmth and colour fidelity of tungsten sources.

Refinements were made to enable the dimming of fluorescent and to improve its colour rendering. But the biggest change for light-fitting designers came with the introduction of a new generation of mini-fluorescent tubes in the early 1980s. Different manufacturers made different shapes, but every shape affirmed a new principle: fluorescent tubes could be fitted into small domestic objects, once the exclusive province of tungsten or tungsten-halogen light sources. Also, as fluorescents give off less heat, these mini tubes could be placed close to such materials as wood or paper, so introducing new possibilities for light-fitting designers.

With the advent of tungsten-halogen in the 1960s, designers were similarly freed from the traditional constraints of Edison's and Swan's invention. Mains-voltage tungsten-halogen bulbs last twice as long as conventional tungsten-filament bulbs without sacrificing any of the colour quality. They are also marginally more energy efficient. Improved performance is due to a regenerative process within the tungsten-halogen bulb. The addition of halogen gas further reduced the migration of tungsten molecules, which interact with halogen and redeposit themselves on the glowing filament. The chemical effect of halogen on glass necessitates a quartz lamp.

But the real revolution in tungsten-halogen lighting has been in low-voltage rather than mains-voltage systems. The 1980s saw the widespread introduction of ultra-slim fittings capable of bearing intense pin-beams of light. Low-voltage tungsten-halogen has completely revised the minimum dimensions required to hold and direct a light source: now tensile wires drawn across a ceiling can constitute a light fitting, so small are low-voltage light sources.

Low voltage has proved itself adept in display and discreet accent lighting, especially in shops. But now it is becoming popular in space-conscious homes. A transformer is needed to convert the electric current from mains to low voltage, something light-fitting designers must take into account.

Other light technologies have momentarily diverted the course of modern lighting design. But neon, high-pressure sodium and metal halide lamps have all proved inappropriate for large-scale domestic use, and, for the most part, twentieth-century lighting has developed against the backdrop of advances in tungsten, tungsten–halogen and fluorescent.

Swiss-made low-voltage Yuma lighting appeared in homes during the late 1980s.

THE TWENTIETH CENTURY

The first design pioneers to take advantage of the invention of electric lighting were the makers of Art Nouveau glassware at the turn of the century. The result was a spectacular body of work between 1890 and 1920 showing an often breathtaking fusion of new technology and new decorative trends. The Art Nouveau designers created glowing lamps using motifs derived from nature – with bases in the form of trees, with shades shaped like butterfly wings or flower petals, and with patterning drawn from plants and insects.

In France, Emile Gallé *and the* Daum brothers *created floor and table lamps as an extension of their work with vases and bowls. In the USA,* Louis Comfort Tiffany, *the son of a leading jeweller who was heavily influenced by the French decorative arts, started producing his own lamps at the Tiffany Glass Company. In particular, he experimented with brilliant colour compositions to create a stained-glass effect. Art Nouveau lamps were also wrought from metal, and surviving examples of the period in glass and metal now command high prices at auction.*

But Art Nouveau was a style which assumed different forms in different countries. In Scotland, the architect and designer Charles Rennie Mackintosh created light fittings for Glasgow

Different forms of Art Nouveau: lights by Charles Rennie Mackintosh (*left*) and Louis Comfort Tiffany.

School of Art in 1907 which revealed a more puritanical, formal and economical expression.

In Austria, the movement took an even more austere, linear approach. The Austrian architect Josef Hoffmann, one of a breakaway group known as the Vienna Secessionists, created light fittings for his best-known architectural design – the Palais Stoclet in Brussels (1905–11) – which revealed in its exploitation of geometrical shapes and repeating patterns, how far the group was removed from the mainstream Art Nouveau.

ARCHITECTS AND LIGHTING

The interest shown in decorative light fittings by architects Hoffmann and Mackintosh in Europe, and by Frank Lloyd Wright in the USA, in the early years of the twentieth century proved a false dawn within the architectural profession.

With the exception of Alvar Aalto, who created a memorable floor lamp in 1956, the chair, not the light fitting, proved to be the ultimate domestic-scale challenge for the architect until an explosion of new lighting technologies and ideas in the 1980s led to a surge of interest. However industrial designers have been more than willing to fill the void.

Architectural interest: (*above*) Hoffmann designs for the Palais Stoclet, in Brussels, 1905–11; (*below*) Frank Lloyd Wright lighting for the Unity Church of Oak Park, Illinois, 1906.

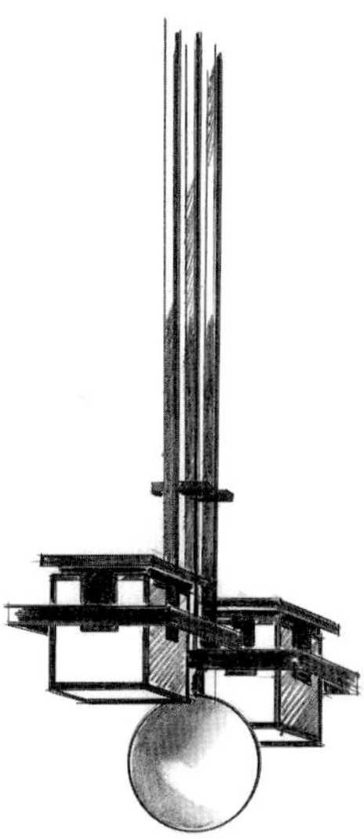

BEHRENS AND UTILITARIANISM

Appropriately, the Godfather of modern industrial design, Peter Behrens, established his reputation with an industrial company which had been set up to manufacture Edison's light bulb. The founder of the giant combine AEG (Allgemeine Elektrizitäts-Gesellschaft), an engineer called Emile Rathenau, saw Edison's achievement at the 1881 Exposition Internationale d'Electricité in Paris and obtained the German manufacturing rights. So began the DEG (Deutsche Edison-Gesellschaft), which became AEG in 1887 when Edison's licence lapsed.

Behrens, an architect-designer who had been a member of the avant-garde Munich Secession, joined AEG as artistic adviser in 1907. He created a coherent identity for the company, so founding the modern corporate-identity business, and went on to stamp his mark on its buildings, prospectuses, catalogues, exhibitions, kettles, electric fans and other products. In particular, his Arc lamp *of 1908 expressed his philosophy of a simple, undecorated utilitarian style. It was the antithesis of the florid, self-conscious Art Nouveau aesthetic he had promoted in his youth.*

At one time or another Behrens employed at AEG such leading names as Le Corbusier, Mies van der Rohe and Walter Gropius, so his influence on the future direction of modern design was immense. Behrens preached the rationalization of objects and in 1907 helped to found the Deutscher Werkbund, an association aimed at uniting art and industry. Certainly the functional turn which light-fitting design was to take owed much to his direction.

THE MODERN MOVEMENT

One of Behrens's assistants, Walter Gropius, designed an exhibition building for the Deutscher Werkbund in 1914 and went on to found the Bauhaus art school in Germany five years later. Between 1919 and 1933 the Bauhaus was the fountainhead of the Modern Movement in design and architecture. Despite having a mere 1,250 students and 35 full-time staff during its 14 years in existence, its influence has been profound on all twentieth-century design teaching. The Bauhaus established the Functionalist design ethic in Europe and when its leading figures fled across the Atlantic to escape Nazi persecution, it did the same in the USA. Most significantly, the Bauhaus was interested in lighting, part of its programme for changing and improving the appearance of all domestic products.

Bauhaus table lamp designed by Karl Trabert, 1930: changing the face of domestic products.

In the Bauhaus metal workshop of tutor Laszlo Moholy-Nagy, students Marianne Brandt *and* Wilhelm Wagenfeld *created a new style of unadorned light fitting with an elegance and beauty deriving directly from its function. In Walter Gropius's own office hung a Constructivist light fitting designed by the Dutch architect* Gerrit Thomas Rietveld, *revealing his links with the avant-garde De Stijl group of artists.*

The reverberations of Bauhaus teaching were felt in Scandinavia, where a Scandinavian Modern movement in design, putting artists and craftsmen in touch with industry, took hold. In the 1920s Danish designers such as Poul Henningsen *and* Kaare Klint *argued the case for change in the provocative journal* Kritsky Revy. *Both Henningsen and Klint went on to create milestone light fittings which showed the accessible face of Danish Modern design, and laid the groundwork for great export success in the 1950s and 1960s.*

Henningsen in particular was a tireless polemicist who dubbed opponents of his theories on soft, indirect lighting 'light's blind book-keepers'. His PH 5 Light *of 1958 remains an ubiquitous classic in Scandinavia.*

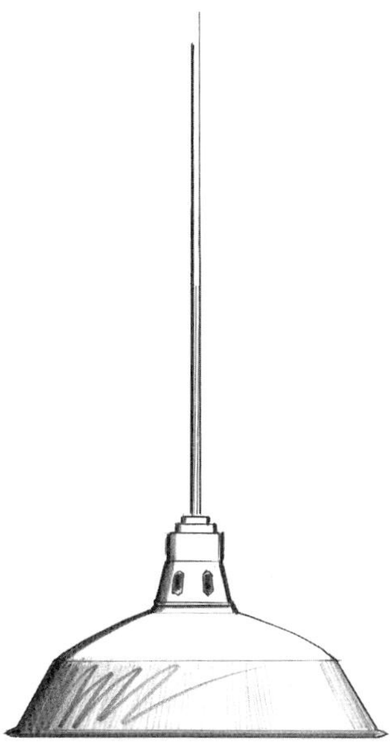

Standard RLM factory light of the 1930s: utilitarian fittings derive form from function.

THE MODERN MOVEMENT IN THE USA

When Walter Gropius and Marcel Breuer fled the Nazis and arrived in the USA in the 1930s, they were able to stamp their brand of European Modernism on the prevailing Machine Age aesthetic movement which had been gathering momentum since the early 1920s.

But the pure design preoccupations of the Bauhaus were at odds with a more commercial school of design founded in the dark days after the 1929 Wall Street Crash by a pioneer group of American design consultants led by Raymond Loewy, Walter Dorwin Teague, Henry Dreyfuss and Norman Bel Geddes. They believed in using streamlining and other styling motifs to appeal to mass markets. Design was a business tool to be exploited, rather than a social and moral instrument for change as Gropius argued.

Art Deco interior of Radio City Music Hall, New York, designed by Donald Deskey.

Light fittings by Walter Dorwin Teague *and his contemporary* Donald Deskey, *designer of the Art Deco Radio City Music Hall interior in New York, show this American preoccupation with 'styling'. Many industrial designers of the 1930s and 40s were known as stylists. But the USA's most popular and widely used desk light from the early 1950s onwards had very different antecedents.*

In the early 1930s George Carwardine, *a young English designer created a desk light based on the constant-tension principles of the human arm. The concept shared in its basis in nature and function a link with both Art Nouveau and the Bauhaus. The result was the famous* Anglepoise *Light which remains in production, scarcely altered, to this day.*

In 1937 Jac Jacobsen, a Norwegian, saw the Anglepoise, bought the patent and renamed the light Luxo. *After enormous success in Scandinavia, he took Luxo to the USA in 1951, where it became, and remains, a common sight in homes, offices and factories everywhere.*

POST-WAR RECONSTRUCTION

Although Britain and the USA saw the invention of the light bulb, and Scandinavia contributed many classic light fittings, attention began to switch elsewhere after World War Two.

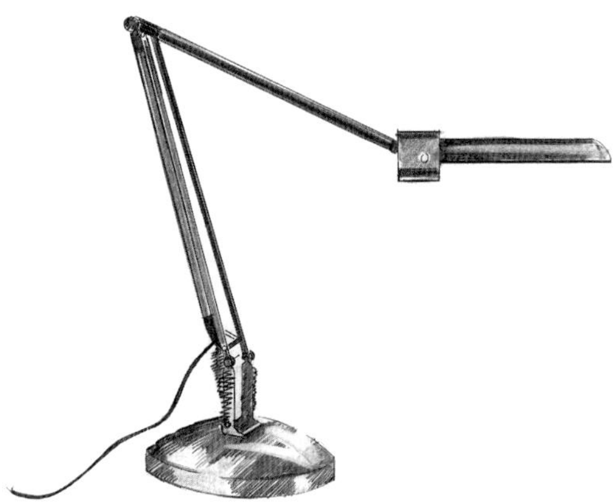

Sleek modern version of the traditional Anglepoise, the world's best-known light.

Italian light fitting of the 1950s: exploring the notion of lamp as industrial sculpture.

In Japan in 1951, the mayor of a small town known for making paper lanterns asked the artist and designer Isamu Noguchi to do something to revive the town's sagging economy. The result was the collapsible paper lampshade, popular around the world today, which diffuses light through its globe shape.

But it was in Italy that the idea of the light fitting as industrial sculpture and 'social metaphor' took hold. After 1945, as Italy struggled to build a new identity in the ruins of Fascism, the ricostruzione, *as it is known, placed industrial design at the heart of economic and social regeneration. The Italians expressed their sense of a new beginning with the most vigorous outpouring of modern design. Domestic products such as furniture, tableware and lighting – low-technology objects which could be rendered unique with a high degree of imagination and style – became the medium by which Italians would signal their new position in world trade.*

ITALIAN DESIGN LEADS THE WORLD

During the 1950s a wide range of novel Italian light fittings became available. Some designs were whimsical, others exaggerated pre-war styles. But Italian lighting firms began to emerge as a major international force. Arteluce, the company founded by the gifted Gino Sarfatti in 1939 and later sold to Flos, won the prestigious Italian Compasso d'Oro industrial design prize in 1955 and 1956.

In 1959 a young missile engineer, Ernesto Gismondi, founded the Artemide lighting company in Milan. Two years later he commissioned a young architect called Vico Magistretti, *who had no experience of designing lights. A totally new and exciting era in lighting design was beginning. Italian light-fitting design would go on to lead the world for the next 30 years or more.*

The bridge between the 1950s and 1960s was built by the Castiglioni brothers, Achille and Pier Giacomo, *who put a*

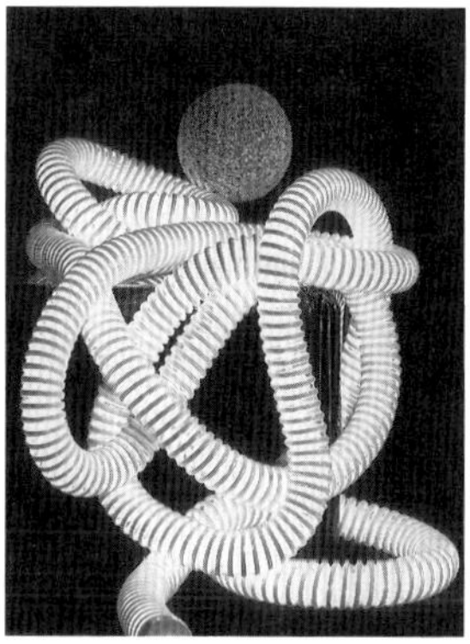

Italian artistry: 1969 plastics Boalum lamp by Livio Castiglioni and Gianfranco Frattini.

car headlamp on the end of a fishing rod and pointed it upwards. The result was the Neo-Functionalist Toio Lamp *for Flos in 1962.*

The Toio Lamp marked a bridge in more ways than one. It signalled the collision of two aesthetics which had run in parallel throughout the century but had rarely met. On one track was domestic lighting, designed to suit the needs of decorated interiors and dating back to the Art Nouveau of Gallé and Tiffany. On the other were rugged, utilitarian fittings, forged in the spirit of Behrens and designed for use in factories and mines, where the need for efficient lighting was greater.

The process of taking lighting innovations directly from offices and factories into the home accelerated. It had started with fluorescent strip lighting and was set to continue in the 1970s and 1980s with spotlight track, uplighters and low-voltage systems.

The Toio Lamp was also the forerunner to the 'salvage' school of design pioneered in Europe in the 1980s by the London-based designer Ron Arad, *in which lighting and furniture would be created from metal and found objects.*

THE INTRODUCTION OF PLASTICS

The 1960s saw a period of great creative ferment as new visual ideas and technologies were tested. Materials technology in particular had a profound effect on design. Glass designs had dominated lamp manufacture in the Art Nouveau period, and metalworking was the preferred option in the time of Behrens and the Bauhaus. Right up to the 1960s glass and metal, especially spun steel, were the almost inevitable materials as bulbs emitted too much heat for much else. But then came the revolution in plastics.

In 1924 thiourea formaldehyde plastics, which could be mottled in any colour, were invented in Britain. Their earliest use was seen on Bandalasta lamp shades made by Brookes and

Ron Arad's halogen clock-light, 1986: a functioning clock face is projected onto the ceiling.

Adams in the 1920s. But it was not until the 1960s that plastics came into their own in light fittings, with the widespread use of acrylic, variously known as Plexiglass, Perspex or Lucite. This had been invented in Britain by Imperial Chemical Industries (ICI) in 1933, and developed during World War Two. Now it was made commercially available in transparent, translucent and opaque forms. The Italians seized the initiative in lighting; Milan became a Mecca for international designers, and, in 1971, the German Richard Sapper *combined Teutonic precision with Italian sensuality to create the classic* Tizio Light *for Artemide.*

THE MEMPHIS GROUP

During the 1970s the light fitting was the symbol of Italian design, emblazoning its style, founded on sophistication and good taste, around the world. But the rules were about to change. By the end of the decade Milan was the centre for such avant-garde groups as Studio Alchimia and a revolution in design was brewing. In 1981 the architect and designer Ettore Sottsass *launched the first Memphis collection of lights, vases and furniture at the Milan Furniture Fair. The effect was shattering. Memphis had turned its back on Milanese good taste in general and sculptural sophistication in lighting in particular. Instead it gloried in gaudy parody – of 1950s Italian cappuccino bars, of Egyptology and rock'n'roll.*

Fantastic, irrational shapes were employed; lights were combined with other pieces of furniture; clashing colour shocked the eyes. The stranglehold of twentieth-century Modernism was broken as Memphis ushered in the colourful and chaotic era of Post-Modernist design.

The effects of Memphis were felt throughout international design, but by the mid-1980s its influence was on the wane and there was a classical, revivalist backlash. Many designs from the early years of the century were revived, and some high-performance modern fittings resembled Art Deco lights.

INNOVATION IN THE 1980s

The 1980s proved fascinating for more progressive reasons. It was the decade when everything finally came together: light source and materials technology advanced dramatically; an eclecticism of visual style asserted itself (due at least in part to Sottsass); and architects began to show the same fierce enthusiasm for the light fitting as industrial designers.

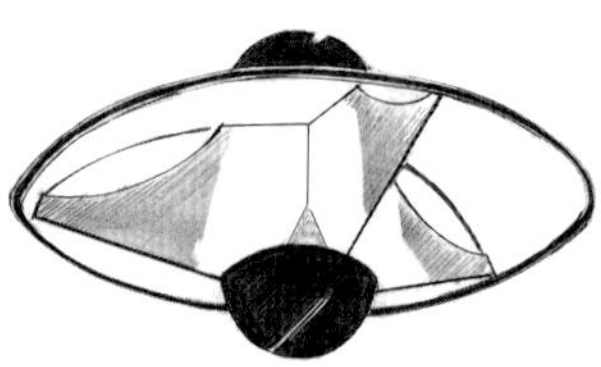

Kite ceiling light by Masayuki Kurokawa for the Japanese manufacturer Yamagiwa, 1987.

The result was a golden era, a decade of radical and often breathless innovation. Italy held poll position on account of the incredible output of its leading firms and the magnetic hold on buyers of Euroluce, its annual September showcase at the Milan Furniture Fair. But many of the highly acclaimed designers in the Italian hothouse came from elsewhere: Ingo Maurer *from West Germany,* Philippe Starck *from France, Masayuki Kurokawa from Japan.*

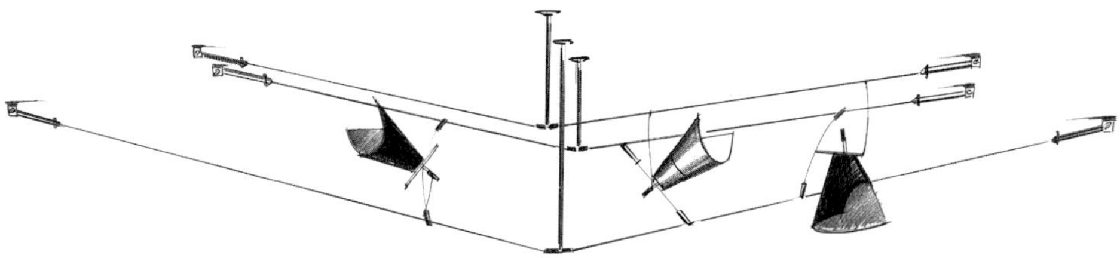

Baka Ru low-voltage lighting system by West German Ingo Maurer shows technical innovation.

Technological innovations proved the driving force: computer controls, touch-sensitive switches and low-voltage and mini-fluorescent sources changed the rules of design. Light sources were cooler so there could be new combinations of metal, plastics, paper and wood in fittings. And smaller so wires suspending low-voltage pinpricks of light across ceilings in the work of Maurer and London-based Shiu-Kay Kan *introduced a new aesthetic. New breakthroughs in structural engineering prompted parallel innovations in the ability of light to span space; for example, the West German firm Erco commissioned two advanced light systems from the architect Roy Fleetwood.*

In fact more innovation in light-fitting design has been crammed into the past ten years than into the preceding five decades, as our selection of milestone designs in this book demonstrates.

Early indications suggest the 1990s will be characterized by a preoccupation with energy efficiency and environmental issues after the style wars of the 1980s. Much work is going on to harness solar power for lighting. Nature may play a more significant role in the lighting of the next century, but we have made a distinctive mark on the lighting of this one.

table lamps

1

The table lamp as a point of focus in a designed interior has changed dramatically during this century. And yet at the same time it has hardly changed at all.

The progression from the Art Nouveau glass lamps of Louis Comfort Tiffany, Emile Gallé and the Daum brothers to the modern industrial sculptures designed by Vico Magistretti and Philippe Starck suggests a vast transformation in technological and aesthetic thinking.

But although new ideologies such as the Bauhaus teaching, new materials such as engineering plastics, and new light sources such as mini-fluorescents have all altered the look and performance of the table lamp, there has been a continuity of function and visual presence throughout this century which explains why so many lights designed in the early years of the modern age remain popular classics today.

LOIE FULLER FIGURAL LAMP

Raoul Larche c. 1900

This dynamic Art Nouveau gilt bronze lamp was designed by the French artist-sculptor Larche in honour of the American dancer Loie Fuller. Light and drapery played a major part in her performances, which captivated Paris at the turn of the century. The lamp, which depicts the dancer waving a swirling robe above her head, represents the rich decorative qualities of Art Nouveau at its most curvilinear.

WISTARIA LAMP

Louis Comfort Tiffany for Tiffany Studios

c. 1900

Tiffany's training as a painter in the New Jersey studio of George Innes and his study of art in Paris are evident in this leaded glass lamp. Its bronze base with circular foot emulates the trunk of a wisteria plant, while the shade simulates the hanging blossom, a design typical of the fascination shown by the Art Nouveau pioneers with the patterns and forms of nature. Tiffany became the most outstanding designer of the Art Nouveau period in the USA, his lamps combining the latest decorative ideas in glassware with the potent new technology offered by electric light.

LILY LIGHT
Louis Comfort Tiffany for Tiffany Studios
c. 1900

The design of the exquisite Art Nouveau Tiffany lamp refines and softens the raw new electric power, featuring 12 glass flower shades which give off an iridescent glow. Each light is decorated inside with a rich gold lustre suspended from its own stem, all twelve being supported on a bronze base embellished with typically naturalistic motifs. The lamp bears the Favrile trademark, denoting Tiffany's handmade range of iridescent and free-shaped glass – a classic among collectors. It was with work such as this that Tiffany won a reputation for excellence at the highest levels of American society: he was commissioned to redecorate the White House and the home of the novelist, Mark Twain.

CAMEO GLASS LAMP

Auguste and Antonin Daum for the Daum factory c. 1900

The technique of cameo glass was first discovered in Roman times but it enjoyed an enthusiastic revival among Art Nouveau glassmakers at the start of this century. The overlaid colours and textures, achieved by working a molten gather of crushed glass, are enhanced by light in a design typical of the Nancy School.

GLASS AND GILT BRONZE LAMP

Loetz c. 1900

Loetz was an Austro-Hungarian company specializing in lustrous iridescent Art Nouveau glass at the turn of the century. This lamp is an outstanding example of Loetz's work: it has a spherical shade supported by four stylized plant stems which enclosed a cylindrical column. The base is gilt bronze.

CAMEO GLASS LAMP

Emile Gallé c. 1900

Gallé was the leader of the Nancy school of design in France, the outstanding Art Nouveau designer, and a great influence on Louis Comfort Tiffany and the Daum brothers. This lamp, with a hemispherical shade above its baluster base, suggests the Japanese influence on European Art Nouveau in its luminous layers of colour.

TABLE LAMP
Donald Deskey for Deskey-Vollmer
1926–7

This Art Deco lamp by American designer Donald Deskey, creator of the interiors for the Radio City Music Hall in New York, suggests the new influences on design in the USA and Europe during the 1920s. Its spirit of Jazz Modernism indicates how Art Deco, successor to Art Nouveau, adapted ideas from Cubism, the Bauhaus and elsewhere to arrive at a popular style. Compared to the Art Nouveau delicacy of Tiffany or Gallé, this lamp is almost crude, with its on-off switch protruding from the base. But, although designed to sit on a table, it hints at being part of a larger scheme.

ETCHED GLASS TABLE LAMP

for Daum 1920s

In its geometrical Art Deco decoration and vibrant sense of colour, this Daum lamp prefigures many of the concerns of Italian design in the 1960s. Acid etching, a technique using wax and hydrofluoric acid first introduced in the mid-nineteenth century, here enjoys a welcome revival. The technique has been found to be suitable for commerical production.

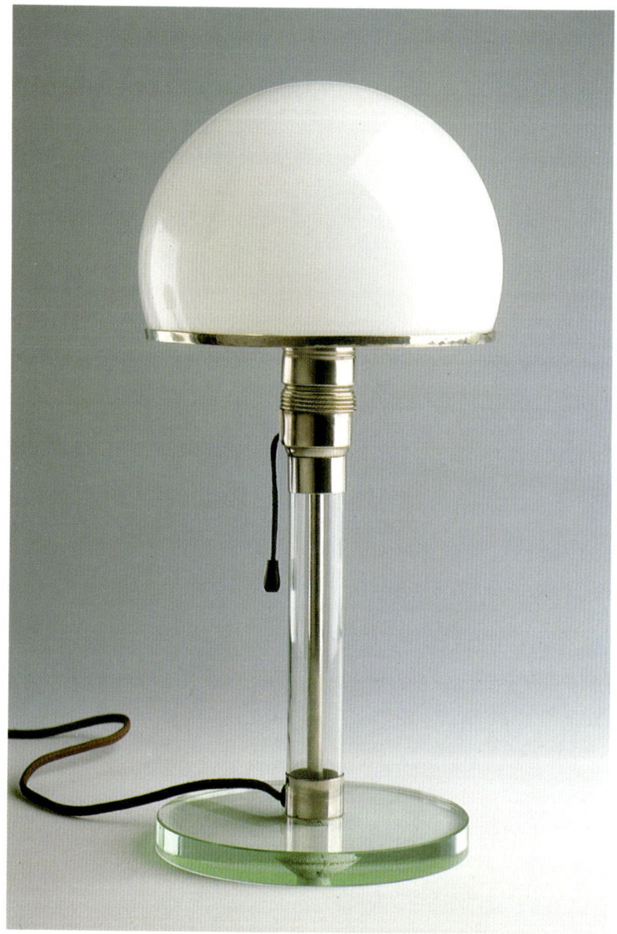

STEEL TABLE LAMP

Wilhelm Wagenfeld for the Bauhaus 1924

One of the most popular and enduring table lamps of the twentieth century, this Bauhaus design by Wilhelm Wagenfeld with C. J. Jucker has been widely imitated. Its beautiful simplicity reflects the purity of the Modern movement as taught by the Bauhaus. In fact, Wagenfeld designed two versions of the chromium-plated steel lamp, one clear and one opaque. Although deriving its imagery from the Machine Age, it has an enduring sensual presence.

BANDALASTA TABLE LAMP

Brookes & Adams 1927

The first introduction of plastics into lamps and lighting, Bandalasta Ware, marketed as 'both artistic and useful', made use of Beetle moulding powders developed in Britain. These enabled the first commercial plastics to offer a colour palette of mottled effects. The Bandalasta lamp gives out a warm glow and is now a rare collector's item.

GLASS TABLE LAMP

René Lalique c. 1925

This outstanding example of the work of the leading French glassmaker Lalique demonstrates fully glass's ability to transmit light in a most imaginative way. A horseshoe-shaped shade with a cherry-blossom pattern moulded in *intaglio* is brought dramatically to life by light shining up from the cylindrical, vertically ribbed body of the lamp it crowns. Despite their bold individuality, almost all of Lalique's designs were mass produced.

NESSO
Gruppo Architetti Urbanisti Citta Nuova for Artemide 1962
This mushroom-shaped table lamp, a modern classic, heralds an era of Italian ingenuity in using newly available plastics to enhance the repertoire of home lighting. The shade is made of cellulose resin; the stem is injection-moulded ABS.

ECLISSE
Vico Magistretti for Artemide 1966
Magistretti's Eclisse table lamp in painted aluminium indicates the fascination among Italian designers of the 1960s with shielding and diffusing light. A revolving screen enables light to be directed to suit individual needs. The circular, zoomorphic shape demonstrates the movement away from the traditional configuration of the table lamp to explore more adventurous forms.

BIAGIO
Tobia Scarpa
for Flos 1968

The Biagio table lamp is one of the most masterly expressions in the 1960s of the Italian light fitting as modern industrial sculpture. It was designed by Tobia Scarpa, who worked initially in the Venetian glass industry. Made from white marble, it gives off a direct light. Its detailing and form are engaging, and it is representative of the high artistic standards achieved in lighting by the Flos manufacturing company.

PANTHELLA LAMP

Verner Panton for Louis Poulsen 1970

One of the most successful results of the collaboration between leading Danish designer Verner Panton and Copenhagen manufacturer Louis Poulsen. Panton set out to create a Modernist interpretation of Art Nouveau. The flowing symmetry of the Panthella lamp shows an instinct for naturalism which is in the best tradition of the Danish Modern Movement.

DALU

Vico Magistretti for Artemide 1970

The Dalu design by Magistretti marks one of the first uses of melamine in a table lamp. This tough, flexible, heat-resistant plastic material gave Italian lighting designers new opportunities to explore geometry and personality when developing fittings. Dalu also demonstrates the continuing sophistication of Artemide, the innovative Italian manufacturing company founded by a former missile engineer, Ernesto Gismondi, in 1959.

ATOLLO
Vico Magistretti
for O Luce 1977

Magistretti's masterly ability to balance aesthetic and functional qualities in table lighting is apparent in this design for Italian producer O Luce. Atollo is made of spun aluminium and finished in baked acrylic. It is widely regarded as a masterpiece on account of its purity of form. All trims and joints are eliminated in a smooth, perfectly precise finish, and the hemispherical shade relates in absolute harmony to the cone base. In 1979 Atollo won a Compasso d'Oro prize for industrial design in Italy and was selected for the New York Museum of Modern Art. The award confirmed Magistretti's reputation as one of the finest light-fittings designers of his generation.

RAY MENTA
Philippe Starck
for XO 1986

This table lamp by France's best-known contemporary designer, Philippe Starck, reflects the technological and stylistic directions of the 1980s. It has a mini-fluorescent light source in its unusually shaped cast-aluminium body, with diffusers to prevent glare. It gives the table lamp an entirely new direction.

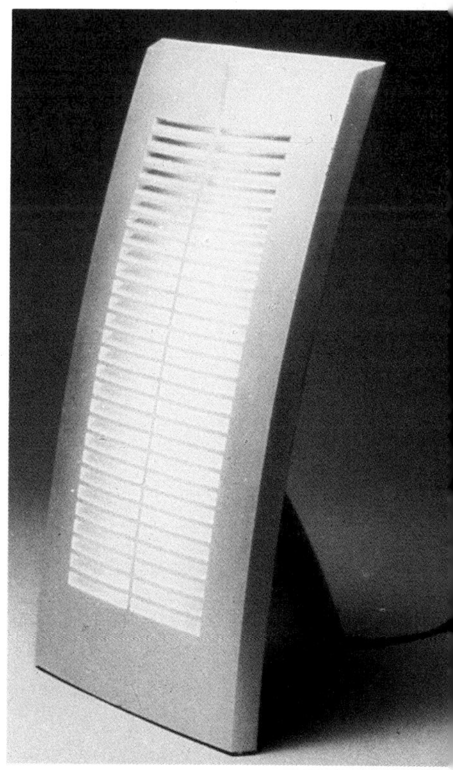

2 ceiling lamps

'A thing is not beautiful until it is correctly formed, and it is not right until it is beautifully formed.' So wrote Poul Henningsen, Danish designer of some of the twentieth-century's most magnificient ceiling lights. That special balance between functional performance and visual presence has been evident in much outstanding work, from the turn-of-the-century tasselled chandeliers of Carlo Bugatti to the Japanese paper globes of Isamu Noguchi in the 1950s and the low-voltage high-wire acts of Ingo Maurer in the 1980s.

Ceiling lights have been the focus for much new design thinking. They have been reshaped by artistic movements, such as De Stijl and the Deutscher Werkbund, but they have also been advanced by technology so that now they can even motor across the ceiling.

CHANDELIER

Carlo Bugatti c. 1900

Eight circular panels in gilt and patinated metal around a porcelain light-bulb mount comprise the structure of this rare chandelier by Carlo Bugatti. Its tasselled fringes recall his interest in Arab culture, evidence of the search for fresh artistic influences for a new century.

CAMEO GLASS HANGING LAMP

Auguste and Antonin Daum for the Daum factory

c. 1900

This patinated iron and cameo glass lamp is typical of the pioneering work carried out by Art Nouveau designers on the decorative potential of central hanging lights. The Daum brothers created a domed shade suspended by three supporting arms from a ceiling rose in the form of a stylized leaf. Each arm has a bell-shaped shade and coloured-glass overlay etched with chrysanthemum patterns. The naturalistic elegance of the design became popular in affluent homes.

ARC LAMP
Peter Behrens
for AEG 1908

Designed by the influential artistic adviser to the giant German AEG industrial company, this lamp is a significant forerunner to the modern generation of undecorated, utilitarian light fittings. Peter Behrens, a founding father of the modern industrial-design profession, rejected the florid Art Nouveau styling of the period by developing a more robust Functionalist aesthetic for AEG lamps. His approach paved the way for later explorations by Bauhaus teachers in Germany and by designers and architects of the international Modern Movement.

HANGING LAMP

Gerrit Thomas Rietveld
1920

This intriguing lamp, a Formalist play on structure in space, is one of the lasting monuments to the avant-garde De Stijl movement in art and architecture. Rietveld, a Dutch architect and furniture designer, brought a practical flair to an essentially theoretical movement when he joined De Stijl. He designed this early example of a lighting system, which shows the influence of Cubism on De Stijl, for his most important architectural patron Mrs Shröder; a version later hung in the office of Bauhaus founder Walter Gropius. The lamp uses standard bulb fittings yet its arrangement on different planes elevates this domestic object to the level of Constructivist art.

LAMPSHADE

Kaare Klint
for Le Klint 1944

Made of scored PVC, this functional and decorative folding shade was designed by Kaare Klint, one of the most influential names in the Danish Modern Movement. It represents the emergence of plastics in Europe in the 1940s as a material to enclose the bare hanging bulb. Le Klint lampshades started production in 1942 under the direction of Tage Klint. Folding paper shades had originally been a family hobby for the Klints.

AKARI PAPER LAMPSHADE

Isamu Noguchi 1950

The standard Japanese-style paper lampshade, a popular feature in countless homes all over the world, dates back to the early 1950s, when the leading Japanese artist Isamu Noguchi was approached by the mayor of a small Japanese town renowned for producing traditional paper lanterns. Asked to develop a light capable of reviving the town's failing economy, he designed the Akari hanging lamp, a collapsible design made of mulberry-bark paper – the paper used by ancient Japanese print artists. The Akari, which means 'light as illumination' in Japanese, proved so inexpensive and attractive – it gives off a warm, diffused glow – that many Japanese manufacturers took up the same idea.

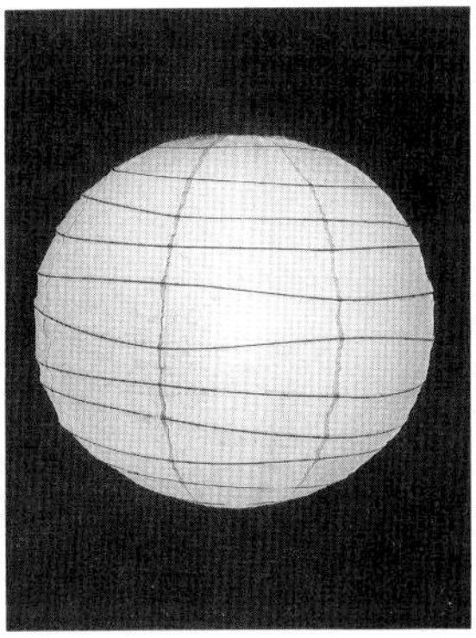

HANGING LAMP

Poul Henningsen for Louis Poulsen 1955

Known commonly as the artichoke light, this magnificent structure represents one of the finest practical realizations of Poul Henningsen's many theories on the art of lighting. Henningsen, Denmark's most famous Modernist lighting designer, was a tireless writer and polemicist on the subject. This design with its overlapping metal flanges, enabled him to achieve a pattern of light and shadow which tamed the raw glare of the bulb and ensured a soft transition between direct and indirect lighting.

PH 5 LIGHT
Poul Henningsen for Louis Poulsen 1958

Henningsen's most famous light fitting adorns interiors all over Scandinavia today. It demonstrates the designer's subtle understanding and confident handling of light. In essence, a simple central globe has been cut and reworked into sculptural segments so that the lamp relies for illumination on reflections from a series of metal planes. Aglow it is a marvellous sight – simple yet so intriguing and effective, the essence of good design.

TARAXACUM
Achille and Pier Castiglioni for Flos 1960

The central globe pendant made popular by Isamu Noguchi in Japan in the 1950s was reinterpreted by two of the most prominent and creative members of the new wave in Italian design in the 1960s. This cocoon-like version of the pendant lamp is made of a metal skeleton covered in a styrene webbing, which is then sprayed with synthetic fibres and layers of vinyl that harden to create a tough outer skin. Taraxacum gives out a soft, diffused light which is very practical. Yet it also makes its own strong visual statement.

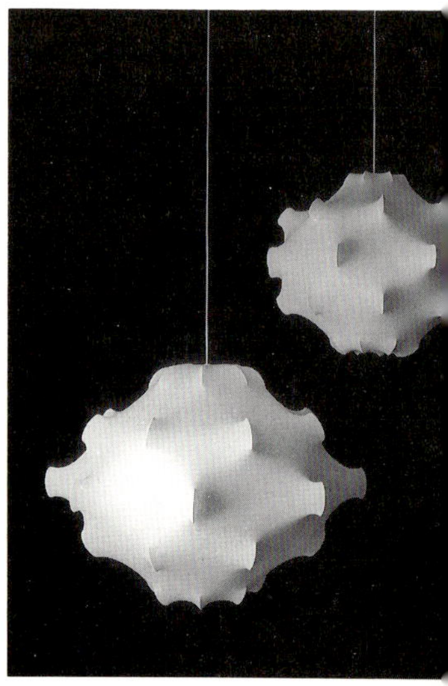

FRISBI
Achille Castiglioni for Flos 1978

This landmark light fitting from Flos shows Achille Castiglioni, the most celebrated of the Castiglioni brothers, reworking the traditional hanging pendant to create fresh visual interest and new patterns of light. Its influential design sends down a narrow shaft of direct light as well as diffusing and reflecting light. A frisbi-shaped acrylic diffuser is supported by three very thin steel wires, and the reflector above is made of chromed metal. The perfect harmony of the fitting's structural elements provides an aesthetic which complements the functional ingenuity of the three-way control of light.

AREA
Mario Bellini
for Artemide 1974

An early example of a flexible lighting system in which the ceiling light is just one element in a modular range of ways to light a room, this design by Mario Bellini features a central suspension pendant on a rise-and-fall mechanism. The light's body is made of polycarbonate plastics; the diffuser, recalling in its form Paolo Venini's 'crumpled handkerchief' Fazzoletti glass vase of 1954, is in a synthetic textile. There are also floor and table-lamp versions of the Area design. Its softer, more sensual approach suggested an alternative to the rectilinear emphasis of much modern design. Once again the Italians were pointing the way forward.

YAMAGIWA LAMP
Verner Panton
for Yamagiwa 1977

Here the lyrical, natural qualities of Danish Modern design have been used to excellent effect for a Japanese company. This ceiling pendant demonstrates Verner Panton's strong interest in organic forms as a basis for the composition of man-made objects. But with the decorative simplicity of the Yamagiwa lamp, he also manages to reflect the essence of early Oriental art. Panton has also conducted valuable research into colour and light.

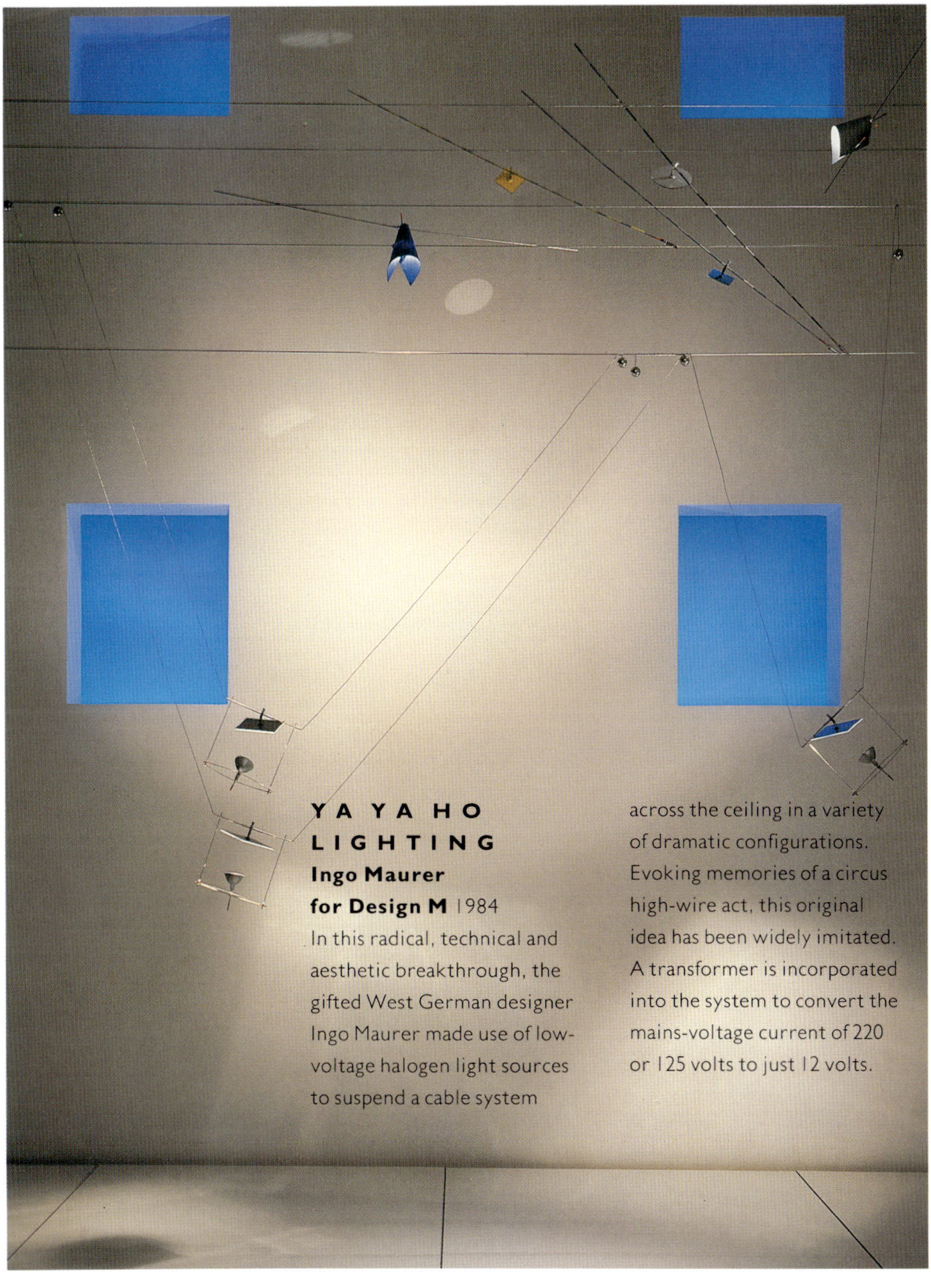

YA YA HO LIGHTING
Ingo Maurer for Design M 1984

In this radical, technical and aesthetic breakthrough, the gifted West German designer Ingo Maurer made use of low-voltage halogen light sources to suspend a cable system across the ceiling in a variety of dramatic configurations. Evoking memories of a circus high-wire act, this original idea has been widely imitated. A transformer is incorporated into the system to convert the mains-voltage current of 220 or 125 volts to just 12 volts.

MIKADO

Ferdinand Alexander Porsche for Artemide 1987

Low-voltage halogen lighting systems grow in sophistication. This ingenious and flexible design by the grandson of the founder of the Porsche automobile empire offers a very wide range of configurations. Electrical contacts in the system are triggered magnetically, simply by placing the aluminium cross-bar extrusions one above the other. A delicately balanced housing encloses each low-voltage light source.

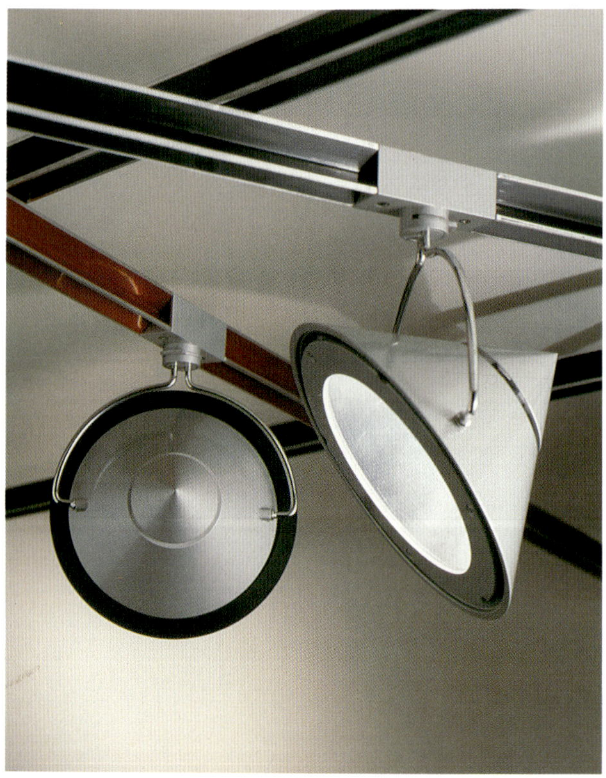

AURORA

Perry King and Santiago Miranda for Arteluce 1983

This pendant lamp skilfully integrates a low-voltage halogen light source into a shallow fitting which consists of a blue acrylic disc sandwiched between two layers of textured glass. Aurora, made in Italy, provides both diffused and direct light, maximizing its light-transmitting qualities.

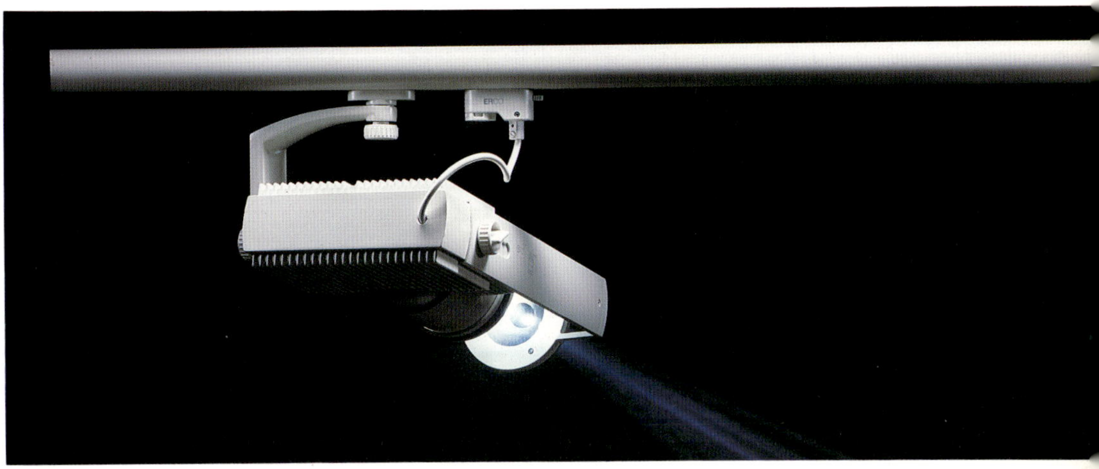

ECLIPSE
Mario Bellini
for Erco Lighting 1986
Eclipse represents a marriage of Italian design flair and West German production excellence in a lighting system which combines elements of camera and stage-lighting technology. It has interchangeable heads to produce different beam widths and solve tricky lighting problems. Although launched as a low-voltage system, the range now includes metal-halide and mains-voltage fittings.

MOTORIZED ROBOTIC LIGHT
Shiu-Kay Kan
for SKK Lighting 1988
The first ceiling light to move across the ceiling was developed in Britain. This remote-controlled, programmable light features low-voltage halogen light sources with built-in reflectors. The android-style unit sits on four wheels and moves back and forth along two horizontal conductive cables.

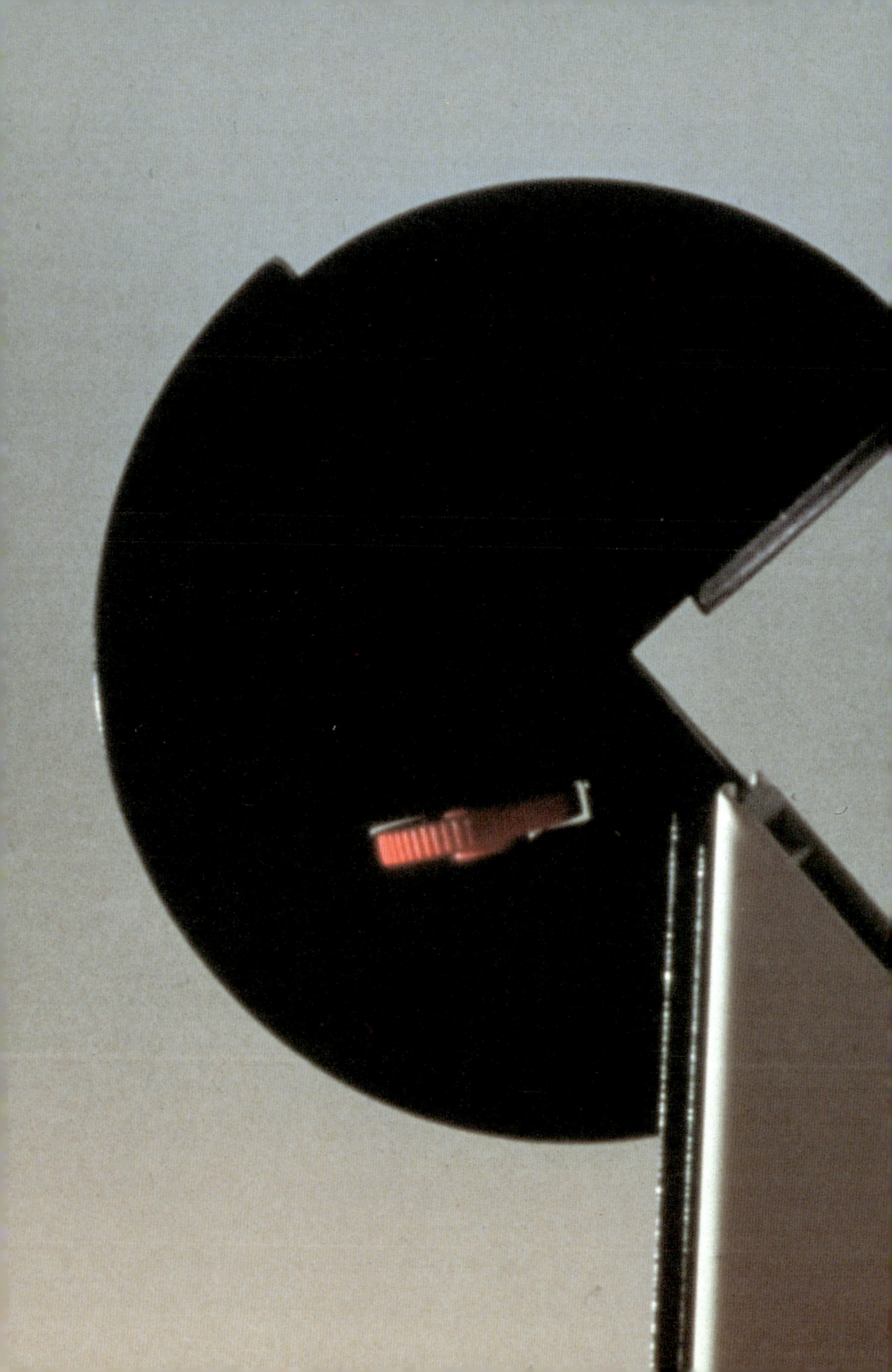

desk lamps

3

Nowhere is the old adage 'form follows function' better demonstrated than in the work or desk lamp which has a specific task to accomplish and must do so with a minimum of fuss. But the robustly utilitarian early task lights, showing evidence of Bauhaus teaching, were also surprisingly beautiful objects to own and use. The cantilever arm of the Anglepoise – the most famous task light of all – has dominated popular taste since it was created in the 1930s, imitated by the Luxo, and updated by the Tizio in the 1970s.

Low-voltage light sources have enabled slimmer, sleeker fittings, many designed by the wonderfully inventive Italians, to take centre stage. But in the end it is functionality which endures, the ability to direct light where it matters – onto the working plane.

BESTLITE
R D Best
for Best & Lloyd 1930

A milestone British-made task light which draws its robust beauty purely from function, this lamp has a universal joint at the head of its arm and hinge-plate which fixes the arm at any point to the base stem to allow for maximum flexibility of height and angle, ensuring that light can be directed efficiently onto the work surface. The shade and base are made of painted spun metal; the tube and joints are chromium plated. The Bestlite has proved an enduring classic: a clear demonstration that an industrial aesthetic can be acceptable in a domestic setting.

KANDEM NIGHT LIGHT
Marianne Brandt
for Korting & Mathieson
1928

This fine example of the functional purity of Bauhaus design is made of lacquered and nickel-plated metal with an adjustable aluminium reflector. Brandt spent time at the Bauhaus in the metal workshop run by the Hungarian designer Laszlo Moholy-Nagy and developed an uncompromising Modernist approach. This lamp successfully set a style which endured for several decades.

ANGLEPOISE LIGHT

George Carwardine for Herbert Terry & Sons 1934

Conceived as an engineering task lamp, the Anglepoise has since become a familiar sight in the home. Carwardine's original brief was for a simple, efficient lamp, assembled from basic engineering components, that could be used at the workbench. He came up with the brilliant idea of a design based on the dimensions and constant-tension jointing principles of the human arm which could be clipped to any surface. It has scarcely altered since.

POLAROID DESK LAMP

Walter Dorwin Teague

1941

Walter Dorwin Teague's angular design for this desk lamp demonstrated the new determination emerging in the USA, to give products a streamlined appearance. Dorwin Teague himself belonged to a select group of pioneering New York industrial-design consultants who were able to attract some of the biggest corporations in American industry precisely because they offered a new way to successfully target consumers – by using design as a commercial tool to appeal to the mass market.

LUXO L-1

Jac Jacobsen for Luxo 1937

Identified by a distinctive grille pattern at the crown of the reflector, this adjustable-arm task lamp owes a great debt to the Anglepoise, although Jac Jacobsen, its Norwegian designer, claimed it was unique. When the Luxo was launched in the USA in 1951 it was an instant and huge success; three years later it was selected by the Illinois Institute of Technology as one of the hundred best products ever made.

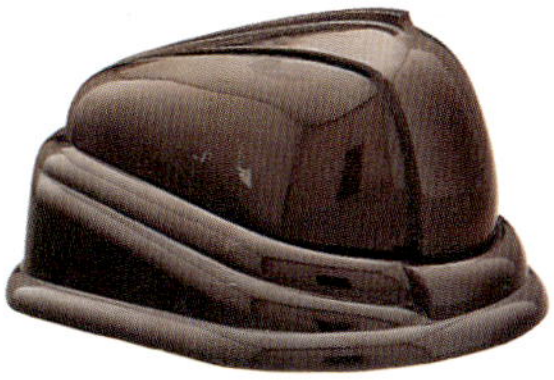

JUMO DESK LAMP

Made in France 1945

This classic streamlined task light has all the personality and presence of an art object. The shade and base are moulded in plastics; the articulated arm is made of chrome and brass. The shade can be angled in several directions while the entire light retracts into the base (*above*). The Jumo lamp, patented and manufactured in France at the end of World War Two, demonstrated that a functional plastics object could make its own artistic statement in the same way that Art Nouveau glass lamps had at the turn of the century. Machine-like forms, synthetic materials and bright colours were set to come of age in the post-war era.

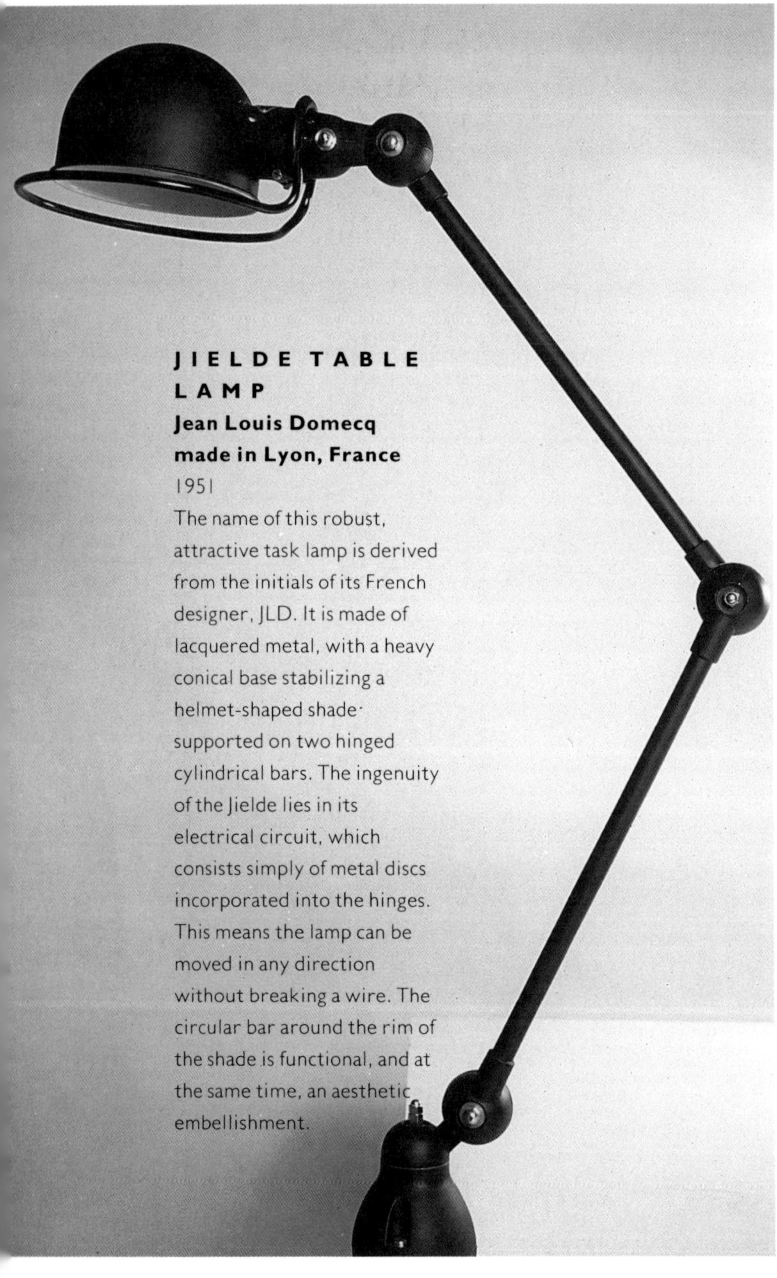

JIELDE TABLE LAMP

Jean Louis Domecq made in Lyon, France 1951

The name of this robust, attractive task lamp is derived from the initials of its French designer, JLD. It is made of lacquered metal, with a heavy conical base stabilizing a helmet-shaped shade supported on two hinged cylindrical bars. The ingenuity of the Jielde lies in its electrical circuit, which consists simply of metal discs incorporated into the hinges. This means the lamp can be moved in any direction without breaking a wire. The circular bar around the rim of the shade is functional, and at the same time, an aesthetic embellishment.

SPIDER

Joe Columbo for O Luce 1965

This much-imitated space-saving design was first created by Joe Colombo, a cult figure on the Italian design scene in the 1960s. Colombo's flirtation with avant-garde art and sculpture ironically produced many highly practical solutions for domestic products. The adjustable table lamp has an enamelled metal lampshade and base and a chromium-plated metal stand. Its compact visual configuration encouraged Italian designers to explore other ways of directing beams of light onto the work surface.

TIZIO LIGHT
Richard Sapper for Artemide 1972

A modern classic, Tizio reworks the adjustable-arm principle of the Anglepoise with great flair and subtlety. It carries electric current to the low-voltage halogen light source along its industrial steel arms so no untidy wires interrupt the stark lines of its taut, economical form. The lamp glides effortlessly into different positions, suggesting Sapper's engineering background as well as his affinity with the sensuality of Milanese design.

GIBLIGIANA

Achille Castiglioni for Flos 1981

A radical departure in task lighting from the grand master of Italian design, Gibligiana has a metal reflector with a mirrored surface; this directs a beam onto the work surface and allows the light to be adjusted and dimmed. The lamp is made in two sizes (standing 41 cm [16¼in] or 52 cm [20½in] high) and has an external body of enamelled metal.

DONALD

Perry King, Santiago Miranda and Gianluigi Arnaldi for Arteluce 1979

In the 1970s Italian lamps began to display wit and whimsy as well as the marriage of function to form. This adjustable desk lamp has a beak-like visor made of ABS plastics which suggests a certain cartoon character, hence its name. Donald has a heavy base of clamp fitting, with an enamelled stem adjusting on a ten-position ratchet mechanism.

SINTESI

Ernesto Gismondi for Artemide 1976

One of the classic task lamps, Sintesi was designed by Italian Ernesto Gismondi, founder of the Artemide lighting company. Made of lacquered metal with an adjustable head, it has a grilled bulb cover which opens up for a greater spread of light. Gismondi, an engineer by training, had no qualms about following in the footsteps of Magistretti and Sapper, two of the star designers he had earlier commissioned.

GOBBO

Marco Zotta for Bitumen 1983

This expressive low-voltage halogen desk lamp represents the creative flowering of Italian lighting design in the 1980s in response to new technology. Tiny light sources allowed for slimmer, sleeker, more versatile fittings and enabled designers to rethink the logic and form of lamps. Sculpted shapes such as the Gobbo light, which echoes the Arco of 1962 in its finely arched all-steel structure, were developed.

NASTRO
Alberto Fraser for Stilnovo 1985

The Italian task light progresses from industrial sculpture to create-your-own-sculpture. A multi-coloured plastics arm containing all the electrical and structural elements can be bent and twisted into any shape required. The result is a highly individual and expressive halogen light fitting which can be instantly customized to suite the task or mood. This brilliant piece of improvisatory design is let down only by a rather clumsy base, housing the transformer, which detracts from the feeling of lightness and flexibility.

BURLINGTON LIGHT
David Morgan for Panasonic Lighting
1986

Modern task lighting comes full circle with a lamp utilizing the latest technology but forging a link with the Art Nouveau glass table lamps of the early 1900s. A dichroic low-voltage halogen light source is placed in the cast-aluminium base of a handblown glass belljar and surrounded by four perforated steel louvres which diffuse some light at the base of the lamp. However, most of the light travels upwards, bouncing off an adjustable mirror at the top and onto the working surface through the glass walls. The design, which evokes the architectural spirit of London's famous Burlington Arcade, won the British Design in Japan competition sponsored by Panasonic Lighting and the Seibu department store.

Wall and floor lamps have been swiftest to be systemized as part of the architectural fabric. Yet they are also points of artistry, sculptural objects within an environment. These conflicting demands have dominated design development during this century, with spectacular individual statements by names such as Louis Comfort Tiffany, Achille Castiglioni and the irrepressible Ettore Sottsass existing alongside more rationalist work by Alvar Aalto, Mario Botta and Giorgio Giugiaro.

Increasingly, wall and floor lamps have incorporated a high level of technology. But that has not deterred designers from achieving extremely creative effects. In the words of Perry King and Santiago Miranda, designers of the famous Jill floor lamp, 'Technology is neutral. It can be loaded with terror or mystery, or it can be made sympathetic.'

wall and floor lamps 4

LUMINATOR

Luciano Baldessari 1929

A magnificent uplighter which appears ahead of its time in its dynamic and sculptural composition, the Luminator is the forerunner to a generation of spectacular and playful Italian lamps, and would not look out of place in the Milanese Memphis collections of the 1980s. Baldessari's artistry evoked the spirit of Art Deco, the circular tube which encircles the main body of the light implying vigour and movement. The Luminator is today reproduced by Luceplan.

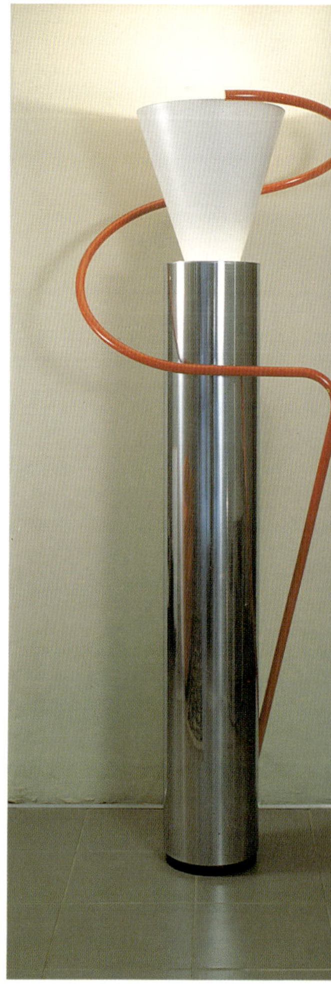

DRAGONFLY FLOOR LAMP

Louis Comfort Tiffany for Tiffany Studios

c. 1900

The conical shade of this leaded glass and bronze floor lamp has a rich decorative quality typical of the Art Nouveau period. But future trends are also suggested by the simple column and unadorned circular base which stands on four feet.

STANDARD LAMP
Alvar Aalto for Artek 1956–8

This lyrical and imposing floor lamp, manufactured by the Finnish company Artek, is a simplified, two-shade version of a light which was originally designed for the interior of La Maison Carré in France. A cluster of three white metal shades were supported by a base and stand clad in black leather (*above*). That version is now an auction-room classic. The design demonstrates Alvar Aalto's attention to the smallest interior details of his buildings and his rationalist, sympathetic use of materials to achieve a gentler, more romantic brand of Modernism. Both versions have since been reproduced.

A R C O
Achille and Pier Giacomo Castiglioni for Flos 1962

The tone was set for Italy's expansive design gestures of the 1960s with this audacious floor lamp by the Castiglioni brothers. A giant arc of slender steel extends telescopically from a block of solid marble to emit direct light from an adjustable polished aluminium reflector with perforations. Arco is on a grand scale: a thrilling statement of light in space which needs plenty of room to work effectively.

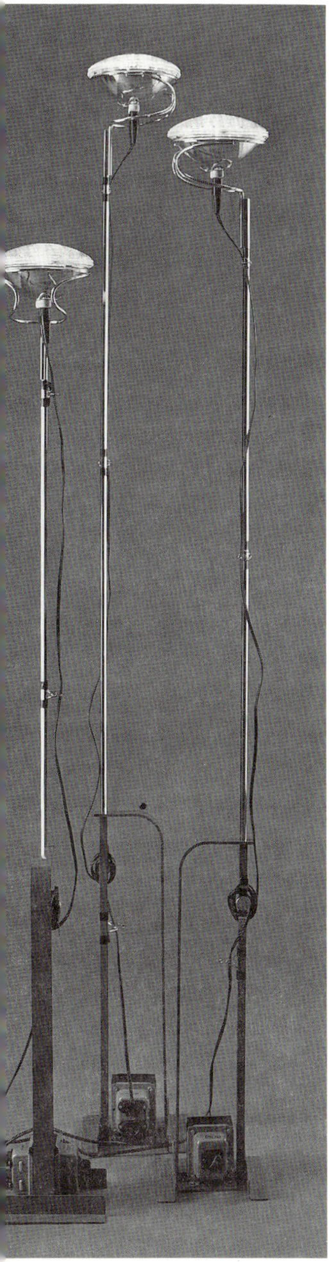

TOIO
Achille and Pier Giacomo Castiglioni for Flos 1962

The industrial aesthetic of salvaged components was made acceptable for the most chic of residential interiors by the tirelessly inventive Castiglioni brothers. In the manner of Marcel Duchamp, they picked at random a fishing rod, band saw and car headlight found lying in a garage and created the Toio floor lamp. This work stands as a milestone in contemporary lighting, introducing a new high-tech design vocabulary to domestic furnishings which has been seized on by successive generations of young designers. Toio remains so fresh and original today that it is hard to believe it is already more than 25 years old.

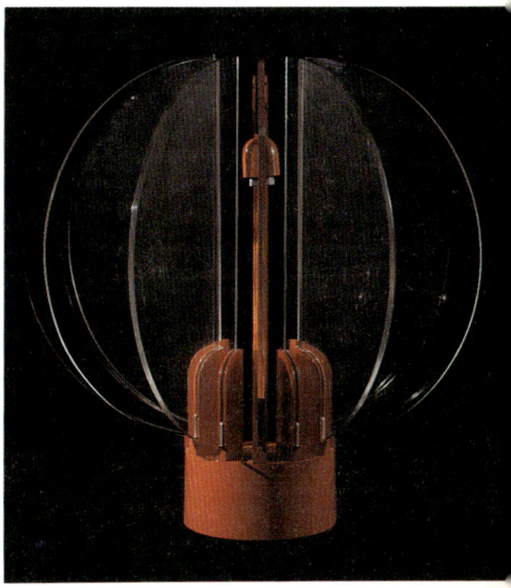

GIRASOLE
Gae Aulenti for Kartell 1968

This ingenious architectural floor lamp exploits the optical properties of clear acrylic by placing a concealed light source inside a structure comprised of nine fan-shaped plastics segments radiating from a central spine and mounted on a varnished aluminium base. The design demonstrates the 1960s' interest in using plastics to create new light forms. The Italian company Kartell led the field in recognizing the potential of plastics, although the prototype for this light was designed for an Olivetti exhibition in Buenos Aires.

COMETA KITE LIGHT

Charles and Jane Dillon

1973

This is the first example of 'soft tech' a style of lighting influenced by Eastern culture. The principle behind Cometa is to enclose the bulb completely with a fabric shade yet maximize the amount of soft, diffused light emitted.

ASTEROIDE

Ettore Sottsass for Poltranova 1968

The influential Ettore Sottsass created a range of floor lamps which were smaller versions of the furniture – cupboards and storage units – he was working on at the time. Asteroide has a wooden base with metal and thermo-formed plastics used for the vividly coloured lamp bodies. This design firmly established the floor lamp as a Pop Art object for the design conscious.

MODELL 265
Paolo Rizzatto
for Arteluce 1973
Rizzatto's robust classic took the task lamp off the desk and attached it to the wall. The design became especially popular in the studios of architects and designers. It is made of enamelled metal and aims for maximum flexibility: the triangular bracket, support stem and perforated reflector are all adjustable. Direct light can be swung to where it is needed.

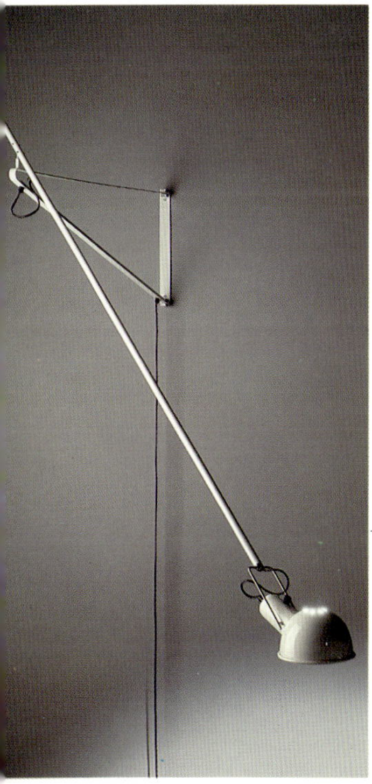

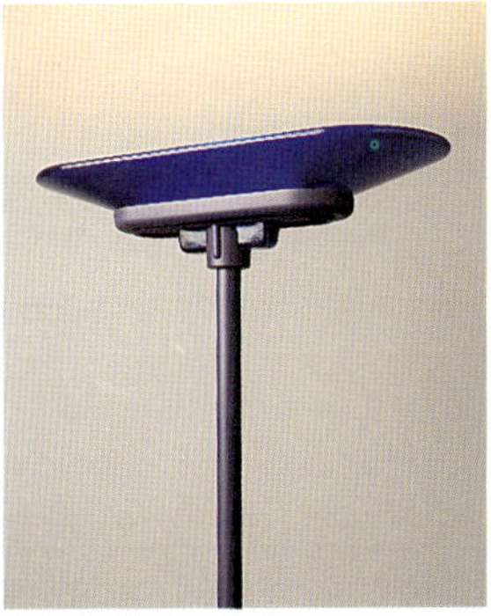

JILL
Perry King, Santiago Miranda and Gianluigi Arnaldi
for Arteluce 1978
This tall and slender Italian floor lamp was one of the first of a new generation of uplighters to make the most of the concentrated power of tungsten-halogen light sources. In fact Jill gives a combination of direct upward light and diffused light, through its etched-glass diffuser (*above*). The circular base, incorporating a dimmer switch, is also made of etched glass. The stem is made of painted metal. The blue glass version of Jill has proved particularly popular. The Milan-based partnership of King Miranda experimented with acid and sandblasting techniques on glass to achieve the effect.

CALLIMACO

Ettore Sottsass for Artemide 1979–80

Sottsass played games with product identity in this aluminium-bodied halogen uplighter which has an integral dimmer. The handle halfway up the stem suggests that the light can be carried like a bazooka. Callimaco was a significant curtain-raiser to the radical rethinking of light fittings in the first Memphis collection, directed by Sottsass in 1981.

MEMPHIS LIGHTS

Ettore Sottsass and Martine Bedine 1981

The light fittings in the first Memphis collection, launched by Ettore Sottsass and his circle of disciples at the 1981 Milan Furniture Fair, indicated the overthrow of the old Modernist order in Italian design. Sottsass reacted against restrained good taste by launching an all-out assault on the senses. Garish colours and wild, irrational forms signalled the new era of Post-Modernism. Sottsass's painted metal Ashoka floor lamp (*right*) symbolized the new approach, its light tendrils waving frantically in all directions. Sottsass also designed the Treetops and Tahiti floor lamps (*far right, top and centre*), but for many people the most amusing and sympathetic light in the collection was Martine Bedine's mobile Super 81 (*far right, below*). This painted steel floor lamp sits on four wheels. The early Memphis pieces were criticized as poorly made but they are now collector's items.

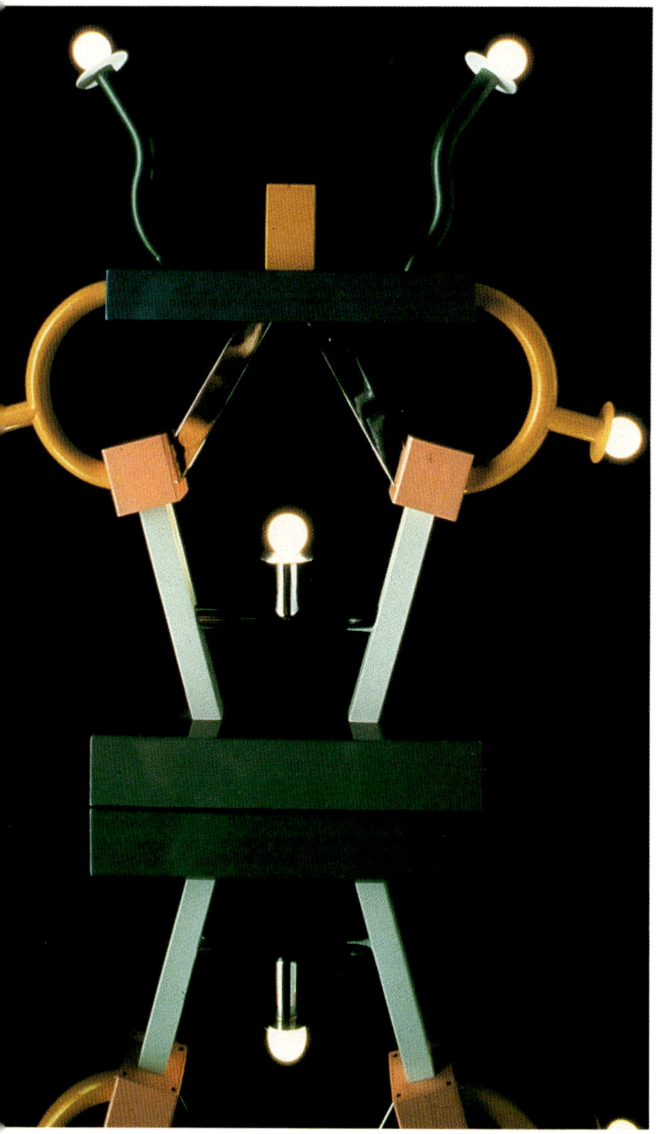

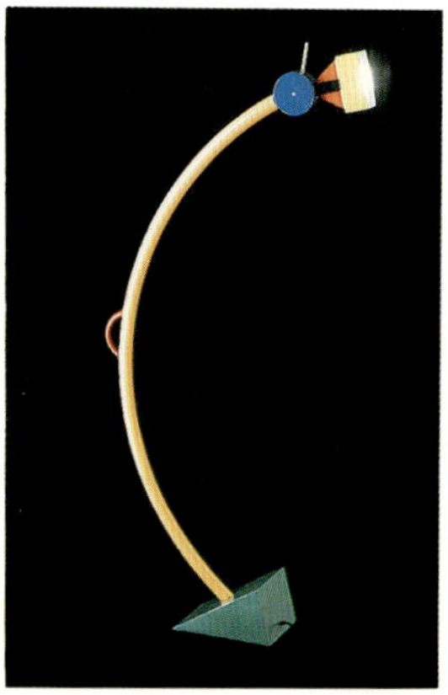

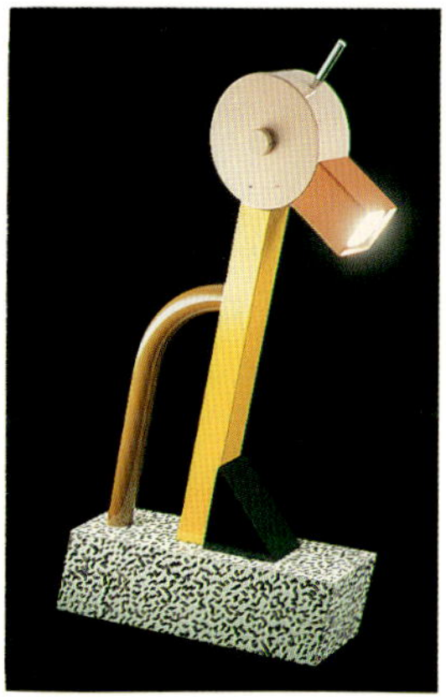

DRAGONFLY LIGHT
Tony Wills
TAG Design 1981

An influential wall-hanging light by British Royal College of Art design-graduate Tony Wills, Dragonfly comprises a folded polyester square shade with two points which slant down spanned by a tungsten-filament light source. The structure is completed by a steel trough with an aluminium reflector. A cable connects the light on its mounting to a wall socket below.

STUDIO LIGHT
Andrée Putman
for Ecart International
1979

The giant black umbrella-style floor lamp is a re-edition by French designer Andrée Putman of an original design which was created early in this century by the famous photographer, fashion innovator and interior designer Mariano Fortuny (1871–1949). Putman's company, Ecart International, specializes in what she describes as 'naive archaelogy', taking neglected design classics and reinterpreting them for modern markets. The lamp operates on the same principle as the photographic umbrella and gives out a soft light via a parabolic reflector.

BUTTERFLY PARETE
Tobia and Afra Scarpa for Flos 1988
This is a wall-mounted version of the Butterfly standard lamp designed by the Scarpas, which uses a pleated fan as a shade. It features a painted steel support and two glass diffusers which shade the bulb and screen out most of the direct light before it reaches the pleated fan. The shade is made of specially treated fireproof fabric. This fitting has a simplicity and directness which conceal its sophisticated technology.

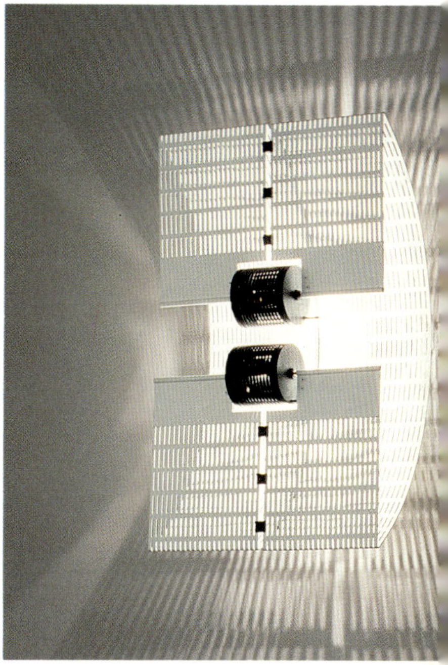

FIDIA
Mario Botta for Artemide 1988
Leading Swiss architect Mario Botta followed in the architectural footsteps of Alvar Aalto and Frank Lloyd Wright by turning his attention to the smaller scale of the light fitting. This halogen wall lamp has an adjustable head and painted perforated plate which casts intriguing patterns on surrounding surfaces.

LUCIFARO
Giorgio Giugiaro
for Luci 1986–7
This adjustable system of spotlights for wall, ceiling or track created by Giorgio Giugiaro, the famous Italian designer of the Fiat Panda and Volkswagen Golf cars, demonstrates the technological sophistication of modern lighting. Made of die-cast aluminium, the system includes both low-voltage and metal-halide versions and incorporates the advanced motifs of contemporary car styling.

BULB BULB
Ingo Maurer
for Design M 1980
A witty Pop Art classic created by West Germany's most inventive lighting designer, the notion of giant light bulb as colourful floor lamp suggests not only the versatility of modern plastics but also the importance of having fun with lighting.

ARIEL LIGHT
Ron Arad
for One Off 1987

The moving floor lamp was developed in Britain by Ron Arad. Its composition from ready-made objects strongly recalls the Toio lamp of 1962 by the Castiglioni brothers. A low-voltage quartz halogen lamp is attached to a telescopic car aerial; jerky movement is achieved via an electric motor and specially printed circuit board. The light is technically clumsy but its spirit nevertheless captures the imagination.

ALVAR AALTO

One of the truly great figures of the Modern Movement, Finnish architect and designer Alvar Aalto (1898–1976) introduced a new visual vocabulary for some of the oldest materials around, such as brick and wood.

His espousal of a gentler, more lyrical and natural brand of Modernism, in contrast to the severe experiments in concrete and steel elsewhere in Europe, has endured throughout this century and is currently enjoying a revival.

Aalto set up an architectural office in Turku in 1927, moving to Helsinki in 1933. He went on to create a series of buildings, interiors, furniture and glassware designs which embodied democratic and humanistic ideals. Unlike other leading architects of the period, his interests extended to lighting and his 1956 floor lamp is now a collector's item.

RON ARAD

Israeli designer Ron Arad was born in Tel Aviv in 1951. He studied at the Jerusalem Academy of Art and the Architectural Association in London, graduating in 1979. His London-based design and retail company, One Off, has built a reputation for often outrageous and bizarre new ideas in furniture and lighting.

Arad takes a highly sculptural approach to batch-production design. His large-scale experiments in concrete and perforated aluminium have created a new Brutalist vocabulary among young designers in Europe. Arad has been described by the architect Richard Rogers as 'a poet of technology'.

PETER BEHRENS

One of the great industrial design pioneers of the early twentieth century, Peter Behrens (1868–1940) was among the first architect-designers to be employed by a large industrial company.

Trained at the art school in Karlsruhe and privately in Düsseldorf and Munich, Behrens went to Berlin in 1907 to take up the post of artistic adviser to the giant AEG (Allgemeine Elektrizitäts-Gesellschaft), a position which he held until World War One began.

He assumed responsibility for AEG's products, buildings and publicity, giving the company its famous corporate identity and overseeing an enormous volume of design work from catalogues and letterheads to electric kettles and lighting. Behrens included among his design assistants such notable names as Walter Gropius, Mies van der Rohe and, very briefly, Le Corbusier. His portfolio of achievements at AEG became the standard by which much subsequent design would be judged.

He became director of architecture at the Vienna Academy in 1922. From 1936 until his death, he was director of the department of architecture at the Prussian Academy in Berlin.

MARIO BELLINI

Leading Italian industrial design consultant Mario Bellini (b. 1935) was among the first designers to realize that the microchip had rendered the old adage of 'form follows function' irrelevant to modern technological products. His work has consequently explored the issues of product identity, often with strongly sculptural forms.

Bellini graduated in architecture from Milan Polytechnic in 1959, and, like his contemporary Ettore Sottsass, created much pioneering design work for Olivetti. Lighting for leading firms – Erco, Flos and Artemide – and furniture for Cassina and B&B Italia also bear his intelligent and tactile hallmark.

Bellini became editor of the influential magazine *Domus* but he is less of a theorist and polemicist than other leading figures in the Milanese design hothouse. He is a highly practical professional, whose achievements were honoured with a major exhibition at New York's Museum of Modern Art in 1987.

MARIO BOTTA

Switzerland's Mario Botta (b. 1943) is an internationally renowned architect of the Rationalist school. He studied

at the Academy of Fine Arts in Milan and then graduated in architecture from the University of Venice.

Botta gained valuable experience in Le Corbusier's studio before establishing his own practice in Lugano in 1969. As well as designing buildings, he has applied his architectural rationale to furniture for Alias and lighting for Artemide.

MARIANNE BRANDT

Marianne Brandt (1893–1983) was one of the best-known designers connected with Laszlo Moholy-Nagy's metal workshop at the Bauhaus. She became a teacher there in 1923.

Although her work largely comprised craft prototypes, several of them were manufactured by Körting and Mathieson. She refined a pure industrial aesthetic which was widely influential in developing a taste for modern domestic design in Europe.

GEORGE CARWARDINE

A British automobile engineer who specialized in suspension systems, George Carwardine (1887–1948) will always be remembered as the designer of the classic **Anglepoise** cantilevered task lamp in the early 1930s. He collaborated closely with British spring-maker Herbert Terry of Redditch to produce a design based on the dimensions and constant-tension principles of the human arm. The result was a model using twisted coil springs which has been scarcely altered to this day.

At the time of the invention of the Anglepoise, Carwardine was one of the directors of the Bath-based firm of Cardine Associates. During World War Two he was involved in secret development work. He died shortly after the war but his legacy – one of the most enduring lights of the twentieth century – lives on.

ACHILLE CASTIGLIONI

The most famous of the three Italian Castiglioni brothers, Achille Castiglioni (b. 1918) graduated in architecture at Milan Polytechnic in 1944 before joining his elder brothers, Livio and Pier Giacomo, in setting up an industrial design studio. Together with Luigi Caccia Dominioni, his brothers had designed a plastic-cased radio for Phonola which was a radical departure in visual terms. This set the new partnership a standard for innovation which Achille Castiglioni was to maintain in a long and highly acclaimed design career.

Early work by the Castiglioni brothers centred on staging exhibitions, an opportunity to experiment with form, colour and light. After 1952 Livio worked independently while Achille and Pier Giacomo remained in partnership until the latter's death in 1968. In the late 1950s and early 1960s, the two brothers collaborated on a series of light fittings for Flos which introduced a new aesthetic to the field.

Achille Castiglioni went on to design products and furniture for such manufacturers as Brion Vega, Zanotta, Knoll and Kartell, winning Italy's prestigious Compasso d'Oro industrial design prize on several occasions. From 1969 to 1977 he was head of the faculty of architecture at Turin Polytechnic.

His work is characterized by wit and invention, and by a restless search for new ways to use ready-made objects. His Mezzadro stool for Zanotta (put into production in 1970, designed with Pier Giacomo in 1954) used a tractor seat; his **Toio Lamp** for Flos (1962) adopted a car headlight.

The Castiglioni brothers came from a family of artists. Achille, one of the great names of modern Italian design, is the last surviving brother.

JOE COLOMBO

A cult figure of Italian radicalism in 1960s design, Joe Colombo (1930–71) studied painting at the Brera in Milan and architecture at Milan Polytechnic. He delved into avant-garde art and sculpture before setting up his own architectural office in 1961.

Colombo's interests swiftly embraced interiors, furniture and light fittings often using moulded plastics. His work in lighting for O Luce is a lasting testament to a rare talent. He died of a heart attack in 1971.

AUGUSTE AND ANTONIN DAUM

Les Frères Daum belonged to the French Art Nouveau movement in decorative glass and a wide range of lamps and vases was produced at their factory in Nancy.

Auguste (1853–1909) and Antonin (1864–1930) Daum were inspired by the success of Emile Gallé at the Paris 1889 Exhibition. They specialized in naturalistic designs, in muted colours, which were acid-etched, enamelled, applied or in cameo relief. From the mid-1920s onwards, the Daum company broadened its variety of decorative styles, embracing the Art Deco movement and even including the work of contemporary artists such as Salvador Dali.

DONALD DESKEY

The designer of the world-famous Art Deco interior of the Radio City Music Hall in New York, Donald Deskey (b. 1894) belongs to that select band of pioneering American design consultants who laid the foundations in the early 1930s of today's international design consultancies.

Along with such contemporaries as Raymond Loewy and Walter Dorwin Teague, Deskey changed the way designers organized their practices and dealt with clients.

Deskey was a leading figure in the Art Deco movement of the 1930s. New materials such as cork and linoleum featured in his design repertoire.

EMILE GALLÉ

The outstanding French designer and glass-maker of his era, Emile Gallé (1846–1904) created a wide and imaginative range of lamps, bowls and vases from 1874 onwards.

His education included studies of both botany and mineralogy. After 1866 he devoted his time to the study of glass-making, travelling to Germany and England. Later he popularized sensual Art Nouveau forms in glassware as enthusiasm ebbed for the bourgeois cut glass of the High Victorian era.

Gallé had great success at several exhibitions towards the end of the nineteenth century, winning prizes for both glass and furniture at the 1900 Paris exhibition. By 1890 he had taken over the direction of his father's factory at Nancy and was able to realize his highly personal interpretation of Art Nouveau ideas. Gallé was fascinated by the decorative motifs of nature – plants, insects and flowers – and by new production methods. Even after his death, the Gallé factory continued to turn out colourful, expressive designs until its closure at the outbreak of war in 1914.

ERNESTO GISMONDI

Italian designer, engineer and industrialist Ernesto Gismondi (b. 1931) was educated as an aeronautical engineer in Milan and Rome. He gained early experience in missile engineering but switched his sights when he founded the Artemide lighting company in 1959.

Under Gismondi's direction, Artemide has become one of the leading names in Italian lighting. Not only has Gismondi commissioned such leading designers as Richard Sapper and Vico Magistretti to design classic lights for Artemide, but he has also successfully produced his own designs.

As Artemide has grown, so Gismondi has emerged as one of the key power-brokers on the Milan design scene. He has invested in other Milanese companies, such as the furniture firm Alias, and he also financed several Memphis collections.

Gismondi is a charismatic and highly unusual industrialist, autocratic in his choice of which designs to bring to market and possessed of a genuine gift for spotting new visual and technical design trends.

POUL HENNINGSEN

The creator of the classic **PH Lamp** now produced by Danish manufacturer Louis Poulsen, Poul Henningsen (1894–1967) was one of Denmark's most innovative and articulate designers.

He did much to bring art and industry together, once writing: 'Throw away your artists' berets and bow-ties and get into overalls. Down with artistic pretentiousness!'

Together with the furniture designer Kaare Klint, Henningsen inspired the Danish Modern design movement. Their joint editorship of the journal *Kritsky Revy* in the 1920s did much to raise popular awareness of design standards, and his lighting designs are today an ubiquitous feature in most Scandinavian homes and offices.

SHIU-KAY KAN

One of Britain's most innovative young lighting designers, Shiu-Kay Kan was born in Hong Kong in 1949. He studied architecture at the Central London Polytechnic and the Architectural Association, and worked with Foster Associates and Fiorucci, before starting SKK Lighting in the mid-1970s.

His first design was the popular Kite Light. Further lighting ideas for home and office followed as Shiu-Kay Kan swiftly explored and exploited new technologies such as low-voltage lighting and motorized, moving fittings.

SKK Lighting has now expanded to include a retail outlet, a production unit developing new fittings, and a consultancy designing contract interior-lighting schemes.

KAARE KLINT

One of the godfathers of Danish Modern design, which enjoyed enormous international success in the 1950s and 1960s, Kaare Klint (1888–1954) was less influenced by the Bauhaus than by the leading English cabinet-makers of the eighteenth century.

Klint is widely regarded as the man who expanded the vocabulary of Danish design by the introduction of unvarnished woods, scored papers and plain fabrics. He established a model of excellence in Danish furniture and lighting which brought much export success.

VICO MAGISTRETTI

One of Italy's most expressive and influential contemporary designers, Vico Magistretti was born in Milan in 1920. A trained architect who began designing furniture and household products in the early 1960s, Magistretti is typical of a generation of architects who switched from town planning to create an imaginative new Italian design style in the era of postwar *ricostruzione*.

Magistretti's collaboration with such leading firms as Artemide, Cassina, Knoll International and B&B Italia has produced many classics in furniture and lighting design. Several of his designs have been acquired for the collection at the Museum of Modern Art in New York.

Magistretti was the man who threw a horse blanket over a sofa to create the Sindbad chair for Cassina in 1982. His innovations with light and form have earned him a place in the Italian design hall of fame.

INGO MAURER

West German lighting designer Ingo Maurer (b. 1932) trained as a typographer and graphic artist before starting his own lighting design firm, Design M, in 1966.

One of Europe's most brilliant and unpredictable talents, Maurer has developed a series of award-winning lighting designs which are much in demand in museums around the world. His gift for new visual, witty and technological ideas was particularly evident in the 1980s.

DAVID MORGAN

British product designer David Morgan (b. 1951) studied physics at London University and industrial design at the Royal College of Art before working as an in-house designer for Thorn Lighting.

Since forming his own consultancy, David Morgan Associates in 1981, he has specialized in lighting products. His **Burlington Desk Lamp** won the 1987 British Design in Japan competition.

VERNER PANTON

Danish furniture, textiles and lighting designer Verner Panton (b. 1926) played a leading role in the explosion of colour and form in the Pop Art of the 1960s.

He started his own firm in 1955, having trained as an architect and worked for Arne Jacobsen, and quickly established a reputation as an innovator by creating the Cardboard House in 1957 and the Plastic House in 1960. His prolific output – from plastic stacking furniture for Herman Miller to textiles for Mira X and lighting for Louis Poulsen and Yamagiwa – has made its mark on the international design scene.

Panton now lives in Switzerland and his explorations of colour theory have led to a greater understanding of the colour spectrum. Recent work has centred on bizarre, organically shaped plastic chairs, and on light fittings which reflect the Post-Modernist spirit of design in the late 1980s.

GERRIT THOMAS RIETVELD

Dutch architect Gerrit Thomas Rietveld (1888–1964) is best known for his contacts with the avant-garde group of artists known as De Stijl.

The legacy of this association includes the Red-Blue Chair of c. 1918, one of the classics of twentieth-century design, and other Formalist experiments with structure in furniture and lighting.

Rietveld's Shröder House in Utrecht, which echoes the ideas of Frank Lloyd Wright, is his lasting monument. But his influence has extended to other branches of design, notably graphics.

RICHARD SAPPER

The designer of the classic **Tizio Light** for Artemide, Richard Sapper (b. 1932) studied mechanical engineering and worked for Mercedes-Benz in Stuttgart before leaving his native West Germany for Milan in 1957.

Sapper worked in the office of Gio Ponti and collaborated with Marco Zanuso on a series of objects which have become high-design cult favourites among collectors. A black television for Brion Vega and a whistling kettle for Alessi are among Sapper's best-known pieces.

He brought to Milan the engineering precision of Germany and married it to the more symbolic and sensual values evident in the best Italian design. Two contradictory strands now dominate his work. One is a studied seriousness – Sapper has been a consultant to IBM since 1981. The other is a wilful playfulness, in which Formalism takes precedence over function.

TOBIA SCARPA

A native of Venice, Tobia Scarpa (b. 1935) followed his father, Carlo Scarpa, into the Italian glass industry before opening an independent design office in 1960 with his wife Afra.

Tobia and Afra Scarpa have jointly developed a highly successful design practice, creating lighting for Flos and furniture for Cassina and B&B Italia.

ETTORE SOTTSASS

One of the most unpredictable, enigmatic and outstanding talents of post-war Italian design, Ettore Sottsass was born in 1917 in Innsbruck, Austria.

Sottsass studied architecture at Turin Polytechnic and opened his own design office in the city in 1946. A pioneering design consultancy to the electronics division of Olivetti, began as far back as 1958.

In the 1960s he became interested in Pop Art and design. In the 1970s he continued his association with Olivetti while carrying out commissions for such firms as Poltranova and Alessi. Sottsass also began producing furniture designs for the avant-garde group Studio Alchimia which he had founded.

But in 1981 he formed a new radical design movement – Memphis. Sottsass unwittingly became the figurehead of the international Post-Modernist movement in interiors and products, even though he protested that the first Memphis collection of lights, chairs and ceramics had been conceived as a joke.

Today Sottsass Associati, the consultancy he formed with architects Aldo Cibic, Matteo Thun and Marco Zanini in 1980, continues to work in the field of industrial design.

PHILIPPE STARCK

A commission by President Mitterand to redesign part of the Elysée Palace propelled French interior and furniture designer Starck (b. 1949) towards international

recognition and stardom.

Starck is now one of the world's top design names. He has designed the Café Costes in Paris, several interiors in Tokyo, and remodelled the New York Royalton Hotel.

Furniture commissions from Disform of Spain, Driade of Italy and others have furthered his scope, and Starck has also put his own distinctive stamp on clocks, cutlery and bottles.

WALTER DORWIN TEAGUE

The son of an itinerant Methodist minister, Walter Dorwin Teague (1883–1960) was one of the USA's pioneering consultant designers who laid the groundwork for the modern design industry.

He studied at the Art Students' League and worked in advertising before opening, in 1926, what he claimed was the world's first industrial design consulting office. Together with New York contemporaries Raymond Loewy, Norman Bel Geddes and Henry Dreyfuss, he established relationships with large US corporations, starting with styling exercises for Kodak.

His legacy is one of legitimizing the role of the designer in industry.

LOUIS COMFORT TIFFANY

Among the USA's most influential and highly acclaimed designers of glass, metalware and jewellery, Louis Comfort Tiffany (1848–1933) enjoyed a reputation for excellence in the Art Nouveau style.

In 1885 he founded his first glass-making factory, Tiffany Glass Company, with his associate Arthur J Nash and began producing decorative vessels and stained glass windows for the luxury market, later adding lamps to the company's repertoire. His work was heavily influenced by French decorative arts after the 1889 Paris Exhibition and he was also acquainted with the English Arts and Crafts movement.

Tiffany's ability to combine Art Nouveau trends in decorative glass and new technology produced a range of table lamps and other light fittings which contributed significantly to the advancement of modern domestic lighting.

WILHELM WAGENFELD

A student of the Bauhaus in Germany, Wilhelm Wagenfeld (b. 1900) has remained committed to the social and aesthetic ideals of the Modern Movement throughout a long career in design and education.

He remained in Germany throughout the rise and fall of the Third Reich. Between 1931 and 1935, he taught at Berlin's Staatliche Kunsthochschule (State College of Art), and saw out the war years at the Lausitz Glassworks.

In 1954 he set up his own studio in Stuttgart to design Modernist utilitarian objects for the home.

Page numbers in *italic* refer to the illustrations